HUMAN-CENTERED

HOW TO DESIGN SYSTEMS THAT ARE BOTH SAFE AND USABLE

Heidi Trost

NEW YORK 2024

"In this much-needed work, Heidi offers a comprehensive exploration of balancing human-centric design and security practices, particularly from the lens of user experience. She delves into the intricate relationship between security, design, and human behavior within a threat-laden digital ecosystem through captivating storytelling. Heidi masterfully unpacks the complexities that arise when trying to create secure systems that are also intuitive and user-friendly. Additionally, she provides practical, actionable strategies to reduce confusion and enhance the balance between usability and security."

—Calvin Nobles, PhD
Portfolio vice president and dean/Human Factors expert

"This is an excellent introduction to human-centered security thinking, and a step-by-step guide to developing security that people can and want to use."

—M. Angela Sasse, PhD
Professor of Human-Centered Security, Ruhr University Bochum

Add this testimonial, please.

"*Human-Centered Security* provides an accessible overview of the user's security ecosystem that demystifies the often overly complex and intimidating world of security and privacy. This book provides clear guideposts and resources for anyone designing for security or with security in mind, which should be everyone."

—Lindsey Wallace, PhD
Director of Design Research and Strategy, Cisco Security

Human-Centered Security
How to Design Systems That Are Both Safe and Usable
By Heidi Trost

Rosenfeld Media
125 Maiden Lane
New York, New York 10038
USA

On the Web: www.rosenfeldmedia.com

Please send errata to: errata@rosenfeldmedia.com

Publisher: Louis Rosenfeld

Managing Editor: Marta Justak

Interior Layout: Danielle Foster

Cover Design: Heads of State

Illustrator: Knox Design Strategy

Indexer: Marilyn Augst

Proofreader: Sue Boshers

HOW TO USE THIS BOOK

Who Should Read This Book?

If you're reading this book, you're probably building digital products.

You're a UX designer or UX researcher or content designer.

Maybe you're a product manager.

You might even be a software engineer or application security engineer.

Increasingly, people who are responsible for security internally at their organization or security awareness programs are thinking about the employee security user experience. If this is you, a human-centered approach to security will help you, too.

What do you all have in common? Whether it's part of your job description or not, you have a massive impact on the security user experience.

You also all care about people—the people who use your products. You want to keep them safe. You don't want them to be hacked, scammed, or manipulated. Obviously, these are terrible user experiences. Often, there are pretty significant financial and legal repercussions for your organization, as well. And, because everything is connected to everything (and everyone) else, there can be a ripple effect of negative consequences for individuals, organizations, communities, and the world.

If you're new to security, you might also be thinking, "But what can I do to help? What is my role in this wild world of security?"

The fact is that if you are designing digital products, you are the one who is responsible for where security impacts the user experience.

What's in This Book?

If you're looking for prescriptive solutions, this book is *not* for you. There are too many ever-changing variables in the security user experience specific to your product and organization that you and your cross-disciplinary team will need to consider. In fact, use a healthy dose of skepticism when you note how other products handle the security user experience. You'd be surprised how many large organizations with lots of resources don't always provide the best security user experience.

Instead, the key is to find what is right for your users and your product. What will help keep them safe?

What this book *will* help you do is the following:

- **Focus on areas of the user experience where security impacts users the most.** (See Chapter 1, "Security Impacts the User Experience" and Chapter 5, "Design for Secure Outcomes.") The good news is that you already have a tool in your toolkit to tackle the security user experience: human-centered design.

- **Understand the dynamics of the security ecosystem.** (See Chapter 2, "The Players in the Security Ecosystem.") Everything in the ecosystem affects everything else. Your product, threat actors, users—they all impact each other and operate within a dynamic ecosystem of technology, geopolitical pressures, life and work pressures, and more.

- **Find your security UX allies.** These are people who can help you improve the security user experience and, with that, security outcomes. While you don't need to be a security expert, you do have to collaborate with your cross-disciplinary teams (product security, engineering, legal, privacy, to name a few) to understand information security risks. Working together helps you understand the full picture and enables you to design more creative, holistic—and more secure—security user experience solutions. (See Chapter 4, "Find the Right People, Ask the Right Questions.")

- **Ask better questions when talking to your cross-disciplinary team.** You'll learn some key security concepts and terms along the way.

- **Tell you what to consider when designing for secure outcomes.** We'll examine some of the most common security user experience issues across products and evaluate how different products have addressed them. (See Chapter 5, "Design for Secure Outcomes" and Chapter 6, "Design Access.")

- **Embrace iteration.** Users will do things you didn't expect or account for. Even more importantly, threat actors will act in ways that you couldn't have predicted. What was effective yesterday might not be as effective today. (See Chapter 7, "Learn and Iterate.")

After reading this book, you'll start to see where security surfaces in the user experience. And you'll feel more confident working with your cross-disciplinary teams to help improve the security user experience.

What Comes with This Book?

This book's companion website (ᴙrosenfeldmedia.com/books/human-centered-security) contains a blog and additional content. The book's diagrams and other illustrations are available under a Creative Commons license (when possible) for you to download and include in your own presentations. You can find these on Flickr at ᴙwww.flickr.com/photos/rosenfeldmedia/sets/.

FREQUENTLY ASKED QUESTIONS

Where does security impact the user experience?

Security impacts the user experience in nearly every part of the user journey. (Check out Chapter 1, "Security Impacts the User Experience," for more details.) Security impacts the user experience most often when a user:

- Signs up or logs in.
- Sets up or configures a device, service, or account for the first time.
- Is asked for personal or financial information.
- Can view or edit the personal information of others (i.e., customer support).
- Receives communications about security or privacy (i.e., an email, a text message, or a security warning related to their device, account, or personal information).
- Has to make a security or privacy decision.
- Has to decide who or what to trust. (For example, is this message/post/website/warning legitimate?)
- Is using a connected device that could influence the physical world (i.e., a car, IoT device, or machinery).

In these scenarios, your users typically aren't thinking about security, which means that you and your cross-disciplinary teams need to be thinking about security.

I'm a designer, so what do I bring to the table when it comes to security? Who will listen to me?

If you design products, I guarantee you are designing for the security user experience—even if you have never thought of it that way before. (Check out Chapter 1, "Security Impacts the User Experience.")

The design decisions you make influence the security (and privacy) choices that users make or the actions they take.

You (and your cross-disciplinary team) understand your product more than your users ever will—including potential security threats that directly impact your users. You are in a unique position to solve for those threats and protect your users from them.

How do I get buy-in to improve the security user experience?

When talking to leadership, try reframing the conversation around trust, rather than focusing on the word "security." *Trust* is where you gain or lose customers. In other words, trust is where the business makes money or loses money.

If people don't trust you with their information, they won't sign up. If they lose trust in you by the way you (mis)handle their information, or if they feel you've violated their safety or privacy, they'll leave. If they can't sign into their account, they'll leave (or rather, ironically, you've made it so they can't come back). Not to mention the thousands of confused and angry customer service messages you'll receive and need to address. (See Chapter 1, "Security Impacts the User Experience" and Chapter 7, "Learn and Iterate.")

When you're looking to get buy-in and promote collaboration with cross-disciplinary teams, take a cue from my colleague, John Robertson, senior principal UX researcher at Secureworks. John actively seeks out different groups at his organization and joins their Slack channels or participates in discussions around the latest research papers on topics like AI and security. John doesn't have to do this—it isn't part of his job description. But, in these low-key forums, the exchanging of ideas is inevitable. John learns about data science and security. The data science and security teams learn about human-centered design. When they have a question, they are more likely to seek John out. And vice versa. Win-win! Do not underestimate these informal channels.[1] Chapter 4, "Find the Right People, Ask the Right Questions," also has more information about finding and collaborating with your cross-disciplinary team members.

---

1 Check out John's podcast episode where he talk more about this: Voice+Code, "What Do You Know About Alert Fatigue? An Interview with John Robertson," *Human-Centered Security* (podcast), July 31, 2024, https://share.transistor.fm/s/30d10844

CONTENTS AT A GLANCE

CONTENTS

FOREWORD

It's not uncommon to hear security-oriented folks talk about how designing secure systems would be easier if people weren't so easy to fool, if they would pay attention to what they're working on. They don't usually say so, but you sometimes get the feeling that my tribe of security people would be happier if there were no people involved at all. Unfortunately, their attitude sometimes crosses the line into contempt.

What's more, if our systems deliver value, our customers will want reliability and resilience. Customers want access to their files, their games, their applications—whatever value your system gives them that needs protection. Customers should be able to trust that their work won't change unexpectedly. Even in a system like Twitter, whose goal seems to be publishing content to the world, there is confidential data involved, like phone numbers and direct messages, and we want messages to come from the real user (or their authorized social media ghostwriter). These are security requirements.

Additionally, we want to protect our business. We want to make sure that we can bill the right people the right amounts for services we can show we delivered to them. These are security requirements.

These requirements don't go away if we show contempt for people trying to use our systems. Customers get harder to satisfy. It makes a lot more sense to make the system easy to use safely. Our customers will be happier.

The day I met Heidi, I had just downloaded a set of "backup authentication codes." You know, the ones that look like this:

Safdaf32-pbmbt   knfdkls-wtafp

And that was, literally, all that was on the page. There was no website or business name. There was no date. There was no URL to go to, in order to use them. It was the security you get when security people and human factors people don't talk.

As engineers, as product designers, and as businesspeople, we need to accept that the world throws complexity at us, and we need to find ways to ship great products anyway. Consequently, we need engineering tools and techniques that enable all of us to address security requirements in human-centered ways.

And that's why I'm so excited about *Human-Centered Security*. This book is focused on human-centered security in a new way. Academic studies of usable security are great, but not enough. This book answers the question: How do we help teams deliver security in a way that works for the people who need it?

Heidi's approach has two elements that resonate for me, and I hope they resonate for you. The first is a focus on the team. Whatever your expertise, none of us is as smart as all of us, and working together, we can deliver better products. If you're coming in as a usability expert, collaboration with security may seem scary. If you're here as a security expert, usability may not have been on your list, but usability really is a security property. The second element is a focus on designing for the properties we want, rather than trying to wedge them in. Getting the right team together early, so the right folks can have input into the decision, helps us make decisions once, and that helps us ship faster, and ship the best product from the first time it's in the hands of the customer.

Adam Shostack
—Author of *Threat Modeling: Designing for Security* and *Threats: What Every Engineer Should Learn from Star Wars*

INTRODUCTION

Whenever I talk to people about the security user experience, the first thing they say is, "Oh...that's really different. Why are you so interested in this space?" Typically, this question is steeped in incredulity. As in, how could someone be interested in something so...uninteresting?

Well, let me tell you.

First, know that I'm not a security expert. I'm just a UX researcher.

As a UX researcher, I'm trained to observe, notice patterns, and inter-pret what they mean for the user experience. And, over the years, I've observed a lot about the security user experience. I observed people ask things like:

How do I know if I can trust this message or website?

I think my account got hacked. What do I do?

What does this security warning mean? Can I just ignore it?

What does [unnecessarily technical term] mean?

How do I know if I set this up securely?

Do you think this message is legitimate? How do I know for sure?

The embarrassing part? I often didn't know how to help them. I can scour help articles and FAQs with the best of them. But many times, I found myself confused, wondering what security terms meant and how to implement the various "best practices" I encountered.

More than a few times, I found myself asking, "Why does this have to be *so difficult*?"

Why do I care so much about the security user experience? Because I care about keeping people safe. And these questions—and the elusive answers to them—are things the user experience can help fix.

That's where you and I—people who build digital products—come in. Much of security happens behind the scenes, where the user is unaware of it. Unfortunately, when security *does* impact users, it's often disruptive and confusing.

As you'll learn in Chapter 1, "Security Impacts the User Experience," security impacts the user in *nearly every part of the user journey.*

These moments are part of the security user experience. UX folks (that's you) have a unique superpower: your job is to understand the people you design for. Turns out, when you understand the people you're designing for better, you can design for more secure outcomes.

You can account for what people know and don't know, where they might make mistakes, or when they have slips and lapses.

You can account for where they might say, "Nope, not today," and create security workarounds.

But...(and this part is important): You can't go it alone. You need to involve your cross-disciplinary teams and proactively talk about security threats.

This will help you and your UX team better understand and anticipate where threat actors will take advantage of humans doing perfectly human things, including where threat actors might trick or manipulate people.

The great news? When you combine superpowers—UX, product security, engineering, legal, privacy (people I refer to as your *security UX allies*)—that's when you have the opportunity to improve the security user experience.

As you read this book, here are a few things to keep in mind:

- When you're in a position to influence behaviors—as you are with the security user experience—question your priorities. Are you truly helping users make the safest choice for them and their circumstances? Check out **Chapter 2**, "The Players in the Security Ecosystem."

- The security user experience is often a mashup of competing priorities. This often leads to tradeoffs where you may be *discouraging* people to take actions that would actually keep them safer. (Go back to the first point: question your priorities.) Further, competing priorities often lead to a terrible user experience for your users. The worst part? It provides a wonderful opportunity for threat actors. Threat actors capitalize on places where users might get confused, frustrated, or create workarounds that are the result of a poor user experience.

- When it comes to the security user experience, there will always be unintended consequences. This book is intended to help you ask better questions. Remember to ask this one: What are the unintended consequences of our security user experience choices?

- But here's a huge disclaimer: I am going to give you a lot of information, but ultimately, there is no "perfect" solution for every product. You need to consider a lot of factors with your security UX allies. Do not tackle the security user experience in isolation—you may do more harm than good. Further, what proves effective today might not be effective tomorrow. That's just how dynamic the security ecosystem is.

While I'm not a security expert, I care about keeping people safe. I know you do, too. Find the people at your organization who can help—your security UX allies. Start asking questions. Together, you can help people stay secure.

Security Impacts the User Experience

Alice is shopping for shoes. She clicks on a link to an e-commerce store. Alice isn't thinking about security. And that's OK, because a lot of the security-related "stuff" happens in the background where Alice doesn't need to worry about it. Until she does.

Because today, instead of a sea of the season's brightly-colored wedges and strappy sandals, Alice, much to her bewilderment, encounters a rather cryptic warning: "Deceptive site ahead."

Alice stops in her tracks, thinking, "Wait...what?"

From Alice's perspective, this message stands between her and her goal: buying the perfect pair of summer sandals.

Alice might question the warning's credibility. She might not fully understand the risks if she proceeds to the site. Alice might question the probability of those risks materializing or how big of an impact they'll have on her.

So, Alice might try to find a way to bypass the warning and continue to the site.

Or she might now distrust the e-commerce website and look to purchase her shoes elsewhere.

Or she might worry that the security of her browser or device might be compromised—but be uncertain of the steps she should take to find out for sure.

None of these scenarios are great user experiences.

In fact, as Jared Spool jokingly says, this user experience—the *security user experience*—actually "SUX."[1]

That's where you—people designing digital experiences—come in.

> **NOTE** SECURITY UX MAKES FOR AN INTERESTING ACRONYM
>
> Jared Spool was the first person I heard use the term *security UX*, and he jokingly uses the acronym *SUX*. I use the term *security UX* because it's clean and neat and aligns with how we talk about other types of UX—enterprise UX, analyst UX, healthcare UX. But my hope is that together we can make sure that security UX no longer "SUX."

---

1   Jared Spool, "Insecure and Unintuitive: How We Need to Fix the UX of Security," UX Immersion: Interactions (presentation), May 2017, www.uie.com/uxsecurity

Imagine a horizontal line—like a timeline—starting from the moment Alice clicked on the link to the e-commerce website and ending where Alice encounters the cryptic warning. Everything below that horizontal line happens in the background, where people like Alice are largely unaware of it. In fact, most users have no idea what happens between their device and the rest of the internet in the fractions of a second between when they click on a link and a webpage appears. These activities simply occur in the background. Some of them are security-related, others are not.

Maybe you, like most users, *also* ignore the security-related stuff that happens below that horizontal line. You think security is something you don't need to worry about because it's someone else's job.

Sure, much of the stuff happening below the horizontal line are things you'll never touch, nor should you.

But you *are* responsible for when the behind-the-scenes stuff bubbles to the surface and crosses the imaginary horizontal line—just like that "Deceptive site ahead" warning did.

In scenarios like these, Alice might be confronted with a choice to be made, an action to take, or she might be tasked with flawlessly recalling from memory some long-forgotten "best practice" that may no longer be effective in today's ever-changing security ecosystem.

The most important takeaway? **When security crosses that imaginary line, it enters your domain—the user experience (Figure 1.1).**

The Odds Aren't in Your Users' Favor

You're reading this book because you care about Alice. You care about your users. You want to keep them safe. But, in order to keep people safe, you must understand the dynamics of the security ecosystem. Unfortunately, the current dynamics don't exactly paint a rosy picture for your users.

Christian Rohrer, vice president of design at TD Bank and formerly vice president and chief design officer at cybersecurity company McAfee, advises designers to think of the three key players in the cybersecurity ecosystem: your users, your product (specifically, where security impacts the user experience), and the threat actors.[2]

---

2   Voice+Code, "How to Design Great User Experiences in a Complicated CyberSecurity Ecosystem with Christian Rohrer," *Human-Centered Security* (podcast), January 6, 2021, https://share.transistor.fm/s/a88f7010

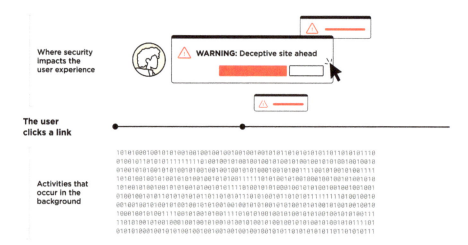

Where security impacts the user experience

WARNING: Deceptive site ahead

The user clicks a link

Activities that occur in the background

```
10101000100101010010010010010010010010101011010101010101101101010111
01001011010101111111101001001001001001010010010010010010010010010
0100101010010101001010010010010010101000100101001111001010010101001111
10101001001010010101001001010010011111101001001001000100100101001010
1010010100100101010010010101010101111010010101000100101010100100100101001001
0100100101011101010101010010110101011110010010010101010111111110101001010
00100100101001010010010010010010010101001010010010010101001001001010010
10001001010010111001001001001001111010101001001001010010010010101001111
110101001010010010010001001001001010010010010010010010010010010011101
010101000100101010010010010010010010010101101010101010101101101010111
```

FIGURE 1.1
When most users click on a link, they are unaware of the complex chain of events—including those related to security—that are happening in the background. That is, until security impacts the user experience.

As Christian explains, Alice often has two aspects of the security ecosystem working against her:

- **The security user experience fails to support Alice in helping her stay safe.** Alice is confused by words and acronyms she doesn't understand and conflicting security "advice." She is expected to understand security and technology concepts, which is unrealistic, and is faced with a series of security roadblocks that are often challenging to navigate.

- **Threat actors are around every corner looking to trick and manipulate Alice and exploit any vulnerability—technical or human—they can find.** Threat actors also know the security user experience is challenging and confusing for Alice, and they use this knowledge to take advantage of Alice.

Christian stresses two more important things to remember about the security ecosystem:

- **All three players are influenced by and influence one another** (as you'll learn more about in Chapters 2, "The Players in the Security Ecosystem," and 3, "Beware of Unintended Consequences"). The organization reacts, which directly impacts the security user experience. The user reacts. The threat actor reacts. And so on.

- **All three players are influenced by the larger systems in which they operate.** Time, money, resources, motivations, technology, organizational structures, power dynamics, new laws and regulations, and geopolitical pressures influence each player in the ecosystem.

Christian's concept of the three players in the security ecosystem, illustrated in Figure 1.2, was so pivotal for me to understand how to improve the security user experience that I used it as the foundation for this book. Alice, threat actors, and where security impacts the user experience (who will be personified as Charlie in just a moment) are characters in the story of the security user experience.

In the following section, you'll get a brief introduction to the three players, but you'll learn much more in Chapter 2.

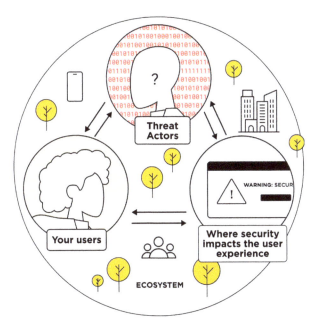

FIGURE 1.2
Everyone and everything in the security ecosystem impacts everyone else.

Alice, Charlie, and the Threat Actor

The first player is Alice, your user[3] (Figure 1.3). Think about how your user would react in a particular situation. One of the things designers do best is to connect the rest of the organization to the real people using their products every day. By using one name through out the book—*Alice*—I'm trying to get you to connect with the human being who is trying to navigate your security user experience.

Alice
💼 Your User

FIGURE 1.3

Meet Alice. Alice is a representation of your user. When you read "Alice," imagine what your users would think or how they would react.

---

3 If you are familiar with security, you might recognize the name "Alice." For example, instead of saying "User A sends an encrypted message to User B," you'd say, "Alice sends an encrypted message to Bob." Whenever you see the name "Alice," think of your end users.

I've chosen to personify where security meets the user experience as "Charlie." No, "Charlie" isn't a persona and doesn't refer to a real person. He's simply a representation of every security message, notification, login or signup screen, or anywhere else security impacts Alice's user experience. When you read the name "Charlie," think about where security crosses that imaginary line and impacts "Alice."

The second player is Charlie. No, Charlie is not a real person or persona. Instead, when I refer to Charlie, think about places in the user experience where security impacts the user. Charlie is the security user experience. Remember the timeline from the beginning of the chapter highlighting where security crosses the imaginary horizontal line and impacts the user experience (Figure 1.1)? That's Charlie. He, or rather it, represents a concept. When you read the word "Charlie," think "where security impacts Alice."

Because many of Charlie's activities happen below the horizontal line, Alice doesn't interact with Charlie all the time. But, when she does, it is often not an experience that Alice enjoys. In fact, Charlie can often materialize as a hodgepodge of competing priorities at your organization and, as such, can be unhelpful, if not downright painful for Alice to interact with. It only seemed fitting to illustrate Charlie as a rather rude, unfriendly, pointing finger directed at Alice (Figure 1.4). You'll learn more about the dynamic between Alice and Charlie in Chapter 2.

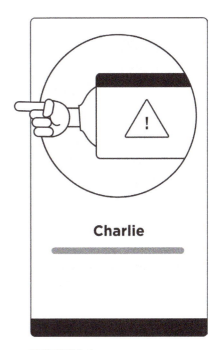

Charlie

FIGURE 1.4
Meet Charlie. Charlie is where security impacts Alice's user experience. Remember, Alice is unaware of many of the things that Charlie does. So, when he does surface and impact the user experience, his sudden appearance can be abrupt, confusing, and, typically, unwelcome.

I found it useful to personify Charlie to help describe the dynamic between Alice, your user, and where security impacts the user experience. Give it a try with your cross-disciplinary team. See if it sticks. You just might hear people say, "Oh, that Charlie is at it again!" That means it's working.

Which brings us to the third player in the ecosystem: the threat actor (Figure 1.5). When I use the term *threat actors*, I'm referring to a person or group of people who take advantage of human or technical vulnerabilities for their own motivations (often those motives are, at

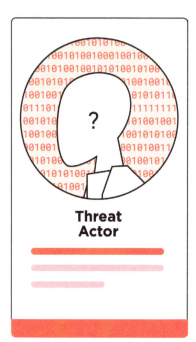

Threat Actor

FIGURE 1.5

Meet the threat actor. Many of the threat actor's actions hide beneath the surface, where Alice is unaware of them. However, the threat actor may try to trick or manipulate Alice, including posing as Charlie.

least in part, financially motivated[4]). Like Charlie, many of the threat actor actions happen below the surface, where Alice is not aware of them.

The threat actor may look for opportunities where Alice made a mistake, such as where she reused a password or misconfigured a database. But the threat actor may also try to trick or manipulate Alice. The threat actor may even pretend to be Charlie in order to trick Alice (for example, by appearing to be a website or dialog box from a product that Alice uses and trusts).

I intentionally didn't give threat actors a name. In many cases, you never know who threat actors are, and you typically don't know much about them. They could be (or sponsored by) a nation-state, in organized crime, an individual, a disgruntled employee, a users' partner who is using technology to stalk, control, and manipulate them . . . or any combination thereof. This book focuses on the broad commonalities of what threat actors are going after and how they go about achieving those ends—in particular, how that might impact your users (see Chapter 2). Your product security team, however, has additional helpful insights as to the specific threat actor motivations and tactics that impact your product and its users. Read more about how to find and collaborate with your cross-disciplinary teammates in Chapter 4, "Find the Right People, Ask the Right Questions."[5]

---

4 For more information, check out the "2023 Data Breach Investigations Report," Verizon, 2023, www.verizon.com/business/resources/reports/2023/dbir/2023-data-breach-investigations-report-dbir.pdf

5 If you are interested in creating personas for threat actors, Adam Shostack helpfully provides some ideas at the end of his book *Threat Modeling*. Adam Shostack, *Threat Modeling: Designing for Security* (Indianapolis, Indiana: John Wiley & Sons, Inc., 2014).

In the security user experience—moments where your product is asking Alice to understand something, make a decision, or take an action related to security or privacy—you are influencing what Alice may or may not do. You may have heard of the term *nudge*, popularized by Richard Thaler and Cass Sunstein in *Nudge: The Final Edition*.[6] Related to UX, nudges are the design decisions you make that influence user behaviors. As the authors point out, as designers you do this—whether you intend to or not—with every single design decision you make. You've made it really easy to keep scrolling on social media (gee, thanks). But you've also made it easy for users to create and save passwords and passkeys using their device (seriously, thank you).

Similarly, you can help Alice understand her options and make the right decision or take the right action for her. Or you can make it more difficult. You could confuse Alice with mysterious acronyms or force her to jump through hoops to do something that is intended to help keep her safe.

But your product isn't the only thing that is influencing Alice. Threat actors—people, or groups of people who take advantage of human or technical vulnerabilities—are also trying to influence Alice. They may, for example, attempt to entice Alice to visit a website or download software that will compromise her device.

---

6   Richard Thaler and Cass Sunstein, *Nudge: The Final Edition* (New York: Penguin Books, 2021).

Where Security Is Most Likely to Impact the User Experience

Imagine that you are a UX designer at a software-as-a-service company (SaaS). Security might impact Alice's user experience at the following moments throughout Alice's user journey (as illustrated in Figure 1.6)

1. Before sign-up.
2. During sign-up.
3. When Alice is asked for personal or financial information.
4. When Alice is configuring security or privacy settings.
5. When Alice has to log in again.
6. When Alice is asked to make a security- or privacy-related decision.
7. When Alice must decide who or what to trust.

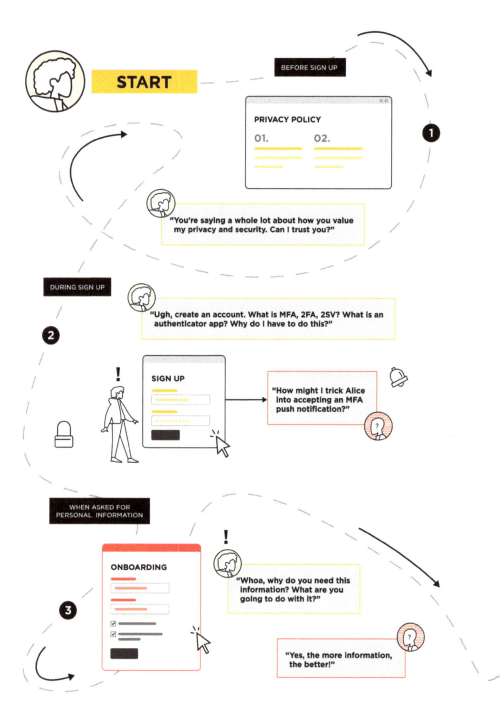

FIGURE 1.6
Security impacts Alice in nearly every part of the user journey.

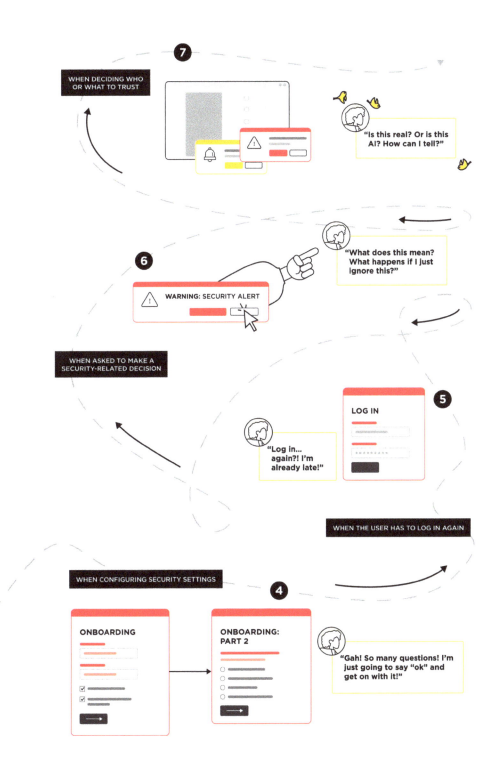

Here's how it might play out.

1. Before Sign-Up

Before Alice decides to sign up to use your product, she checks out your marketing website, evaluating whether your product can help her.

She pauses. "You're saying a whole lot about how you value my privacy and security. What does this mean? Should I be concerned?"

Words like *security* and *privacy* (including your privacy policy) on your marketing website—that is Charlie surfacing. And Charlie is causing Alice to question her decision to sign up for your product. What Alice is really asking in this moment is, "Can I trust you?"

2. During Sign-Up

Now Alice is at the sign-up screen (Charlie again). She sighs, "Ugh, create an account. That stinks. Just another password I have to remember."

Alice chooses a password she typically uses on a few other accounts. But she gets an error message that the password is not strong enough (Charlie again). Alice thinks, "This password was 'strong' according to the other websites I use it on. Why are you telling me that it isn't strong enough? This is really irritating."

Now Alice needs to set up multifactor authentication (MFA). MFA means that Alice will need to use two or more authentication methods to sign into her account. In addition to her password, she might use a time-sensitive code sent via text or insert a hardware security key (like a YubiKey) into her device. You'll learn more about MFA and walk through an example in greater detail in Chapter 3.

Confusingly, Alice might see different acronyms across the different services she uses: MFA, 2FA, 2SV. While there are differences between these, the nuances are something Alice is unlikely (and shouldn't be required) to know. Alice may ask, "What do these acronyms mean: MFA, 2FA, and 2SV? Is this the same thing I do to sign into my bank? Or is it different?"

She might also think, "Why do I have to do this? What are the steps involved? And what is an authenticator app? Where do I get one, and how do I use it?"

The threat actors, of course, *love* Alice's confusion. They prey on the fact that Alice might not set up MFA because she doesn't know what it is, or they'll find ways to trick Alice into giving out her MFA codes or bombard her with push notifications, hoping she chooses "yes" just once.

3. When Asked for Personal or Financial Information

When Alice is asked for personal or financial information, she balks. "Whoa, why do you need this information? What are you going to do with it?"

If Alice's information can be seen by others, such as on a social media site, threat actors may use Alice's personal information as intel to be used against Alice. For example, to send her convincing phishing messages, to trick or manipulate Alice, or to stalk, threaten, or extort Alice.

4. When Configuring Security and Privacy Settings

Later in the onboarding process, Alice might be asked to configure some security and privacy settings. If this SaaS product is for her business, she might even be adding, deleting, and managing other users. Often, Charlie is unhelpful during this process, forcing Alice to muddle through a confusing, multistep process by herself.

Alice might think, "Gah! So many questions! I'm sick of reading and clicking. I'm just going to say "OK" or "skip" without reading all of this and get on with it."

And, you guessed it, threat actors prey on this, as well. They know Alice went with the system's default security settings without knowing what other options were available to her, forgot to change a default password, or misconfigured the product by accident. To be clear, this is typically through no fault of Alice's. Many of these issues could be prevented if Charlie anticipated what Alice needed and helped her through the process.

> **NOTE** DECISIONS THE USER MAKES DURING SIGN-UP AND SETUP CAN IMPACT A USER'S SECURITY LATER
>
> Decisions that the user makes during the earlier stages of the user journey (such as choosing a strong password or enabling multifactor authentication) can impact their experience later in the user journey (for example, if their account is hacked or they lose the phone they use for multifactor authentication).

5. When the User Has to Log In Again

Eventually, Alice needs to log in again or reauthenticate. Charlie again. She thinks, "Log in…again?! I have to do this really important thing, and I'm already late!"

Maybe Alice tries to log in and can't get access. She might think, "Why am I locked out? I swear this is the right password. Is there someone who can help me? Anyone…anyone?" Hm, all of a sudden, Charlie is nowhere to be found.

One day, Alice gets a message that someone has signed into her account, but it wasn't Alice. Confused and somewhat alarmed, Alice thinks, "Is this message real? What does this mean? What should I do (and do I really have to do anything)?"

Is this Charlie? Or is this a threat actor posing as Charlie? How is Alice to know?

6. When Asked to Make a Security- or Privacy-Related Decision

As Alice continues her user journey, she may be asked to make a security- or privacy-related decision or take a security- or privacy-related action. The browser Mozilla Firefox, for example, helps users by warning them when it suspects a downloaded file may be harmful. Firefox surfaces the warning, shown in Figure 1.7,[7] giving Alice an opportunity to pause, consider, and decide if she wants to proceed with downloading the file or not.

FIGURE 1.7
Mozilla's Firefox browser will warn users if it detects a file they've downloaded might be harmful, as shown on a Mozilla Firefox support page.

---

7   "How Does Built-In Phishing and Malware Protection Work?" Firefox Support, updated January 12, 2024, https://support.mozilla.org/en-US/kb/how-does-phishing-and-malware-protection-work

1Password, a password manager, helps users manage their account details—like usernames, passwords, and passkeys—across the myriad of digital services people use every day. Chances are, at some point, one of the hundreds of websites and services that people use will be impacted by a security breach. 1Password knows that users may not be aware when this happens (after all, who can possibly keep track of the hundreds of services they use every day?). Because of this, 1Password alerts users when a website they use was impacted by a security breach and gives them a specific action: go change your password on the affected account, as shown in Figure 1.8.[8]

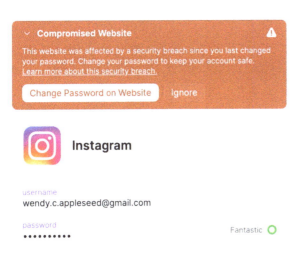

FIGURE 1.8
1Password does a great job of identifying and offering solutions for potential issues users may have with the many accounts users must keep track of, as shown on a 1Password support page. In this case, 1Password encourages the user to change their password on a website that had a security breach.

When asking Alice to make a security- or privacy-related decision, keep this in mind: Alice may question the meaning and legitimacy of the communication. She may ask, "What does it mean that this website/file/person is a 'potential security risk'? How does this impact me? What if it's not dangerous, and I don't respond or I miss out? What happens if I just ignore this?"

8 "Use Watchtower to Find Account Details You Need to Change," 1Password Support, https://support.1password.com/watchtower/?ios

Because of this, when and how you surface security-related information and warnings really matters. The words you use are critical and will influence what Alice does next. Check out Chapter 5, "Design for Secure Outcomes," for more information.

7. When Deciding Who or What to Trust

Imagine that Alice is interacting with other users on your platform. If you have a social media platform, Alice might question whether the content she is viewing is generated by AI or might be an attempt to manipulate her. She may look to your product to help her distinguish AI-generated or potentially manipulative or deceptive content (Figure 1.9)—a task that Alice will not be able to do without help.

Similarly, Alice may question whether legitimate communications coming from your organization—emails, phone calls, texts, push notifications, and more—are, indeed, originating from your organization or are part of a deceptive threat actor ploy. The more noise Alice has to sift through, the more confusing it is for her.

FIGURE 1.9
Social media platforms are working on ways to help users identify whether content, such as photos and videos, were generated by AI, such as putting labels on an AI-generated photo.

What is powerful about creating visualizations like the chaotic user journey, depicted in Figure 1.6, is that it helps your team think from the user's perspective. When the user thinks, "Hmm, I don't really understand what this means," they may leave and never come back. Or they might ignore a message that was intended to help them or prevent them from being put in harm's way. Or they may not set up your product in a way that could help keep them and their organization stay more secure.

> **NOTE** DON'T THINK ABOUT SECURITY IN ISOLATION—
> IT'S WOVEN INTO THE ENTIRE USER EXPERIENCE
>
> Keep in mind, the user experience of security is interwoven as part of the overall user experience. It's not just one moment—it is many moments. As with any user experience, it's critical to think holistically, even if you are designing for a specific workflow.

Human-Centered Design Becomes Human-Centered Security

Know that you already have a very important tool in your toolkit: human-centered design.[9] Human-centered design can be applied to any domain, including security. And security is ultimately about keeping people safe. It's *human-centered* security.

---

9 The most important aspects of human-centered design applied to security are those defined by ISO 9241-210:2019, *Ergonomics of Human-System Interaction, Part 210: Human-Centred Design for Interactive Systems*. The process is iterative, multidisciplinary, and emphasizes understanding and involving users as part of the design process. In addition, the larger system, or environment, is taken into account.

ISO 9241-210:2019, *Ergonomics of Human-System Interaction, Part 210: Human-Centred Design for Interactive Systems*, 2nd ed. (ISO, 2019).

The realization of a need for human-centered security isn't new. Anne Adams and M. Angela Sasse explained the need for user-centered design in security in 1999.

Anne Adams and M. Angela Sasse, "Users Are Not the Enemy," *Communications of the ACM*, 42, no. 12 (1999): 40–46, https://doi.org/10.1145/322796.322806.

Zurko and Simon introduced "user-centered security" in Mary Ellen Zurko and Richard T. Simon, "User-Centered Security," *NSPW '96: Proceedings of the 1996 Workshop on New Security Paradigms* (September 1996), 27–33, https://doi.org/10.1145/304851.304859

Remember, where Charlie—and threat actors—impact Alice is *your* responsibility. When you see Alice struggle (or anticipate she will), don't stay silent. Document your concerns. Will you have to learn new words? Yes. But haven't you learned a new vocabulary unique to every new domain you have worked in? I know I have. Don't be afraid to ask questions.

You learned that Alice has a lot stacked against her. The security ecosystem is a gnarly one—and it's imperative that you understand and account for it. You'll need to lean into systems thinking now more than ever when designing for the security user experience.

But there are two other components of human-centered design you need to embrace:

- Cross-disciplinary collaboration
- Iteration

Cross-Disciplinary Collaboration

You need to involve stakeholders (I refer to them as your *security UX allies*) not only from product and engineering, but also from product security, legal, compliance, and privacy teams. While you can help them better understand your users, they can help you understand what regulations you need to consider, which security threats are specific to your product and use cases, and where threat actors may try to trick or manipulate your users. They know the technology limitations—and its strengths.

Let me be clear: nothing in this book will help you improve the security user experience without their help. In fact, without them, you might do more harm than good. You'll learn more about finding and collaborating with different people at your organization to improve security outcomes in Chapter 4.

Embrace Iteration

Question assumptions. You know you won't get things right the first time. So, you should build in iteration as part of your design process. For the security user experience, users and threat actors will do things that you could not have anticipated. An iterative process is required to address the dynamic security ecosystem in which your product exists. You'll learn more about iteration in Chapter 7, "Learn and Iterate."

But let's pause for a moment and define what security actually means.

Develop a Shared Definition of Security

By now, you're probably thinking that you've just about finished the first chapter of *Human-Centered Security*, and you still haven't been given a definition of security. I'm first going to give you the technical definition, and then I'm going to argue why the best definition of security is the one your cross-disciplinary team defines together.

Why? When I talk to most users, it becomes apparent pretty quickly that how I think about security is not the same way as how they think about it. And they certainly don't think about security the same way that a security expert does. One of the reasons (certainly not the only reason, as you'll learn in Chapter 2) that Alice and Charlie butt heads is because they think about security differently. You and your cross-disciplinary teams need to bridge that gap.

Throughout this book, when I say *security*, I'm referring to *protecting information and information systems*.

In "Guide for Conducting Risk Assessments," the National Institute of Standards and Technology (NIST) explains, "Information systems are subject to serious *threats* that can have adverse effects on organizational operations and assets, individuals, other organizations, and the Nation by exploiting both known and unknown *vulnerabilities* to compromise the confidentiality, integrity, or availability of the information being processed, stored, or transmitted by those systems."[10]

The words that NIST uses ("threats," "vulnerabilities," "confidentiality," "integrity," and "availability") are all really important for you to understand what information security means—and will help you have more productive conversations with your cross-disciplinary teammates. More on what those words mean in just a moment. First, let's take a step back to put a more human wrapping around the word *security*.

Kelly Shortridge, co-author of *Security Chaos Engineering: Sustaining Resilience in Software and Systems* and senior principal engineer at Fastly, argues that security practitioners should take a more human approach to security, starting with the very definition of "security." She writes, "We should aim to provide subjective security, that

---

10 Joint Task Force Transformation Initiative, *Guide for Conducting Risk Assessments*, NIST Special Publication 800-30 Revision 1, (Gaithersburg, MD: National Institute of Standards and Technology, 2012), 1, https://doi.org/10.6028/NIST.SP.800-30r1

ancient-school version of *securitas,* which meant freedom from anxiety, fear, or care."[11]

Michael Snell, UX research lead at JPMorgan Chase, said something similar that has stuck with me. Michael says, when team members say, "'more secure and more private' or 'less creepy' what does that actually mean?"[12] And what does it mean *for your users*?

What Kelly and Michael are suggesting, in part, is to develop a shared understanding of security at your organization. In other words, when you talk to people at your organization, you should be talking about what security and privacy mean *for your product* and *your users.*

Asking questions like "What does security mean for our product and our users?" might sound simplistic, but it's imperative to improve the security user experience. What works for your organization, product, and its users might not work for another. See Table 1.1.

If you're looking for a prescriptive solution—there isn't one. And the solution (or whatever was hacked together) that works today, simply won't work tomorrow. That's what makes working in the security field fun! Instead, what I'm hoping to help you and your team do is ask better questions. Most importantly, ask better questions with your cross-disciplinary team.

NOTE WHAT DOES PRIVACY MEAN?

According to *The Privacy Engineer's Manifesto* by Michelle Dennedy, Jonathan Fox, and Tom Finneran, "Data privacy may be defined as the authorized, fair, and legitimate processing of personal information [...] Processing personal information includes, but is not limited to, collection, storage, use, sharing, organization, display, recording, alignment, combination, disclosure by transmission, copying, consultation, erasure, destruction, and alteration of personally identifiable information and any data related to it."[13]

---

11 Kelly Shortridge, "When We Say 'Security,' What Do We Mean?" *kellyshortridge.com* (blog), October 26, 2023, kellyshortridge.com/blog/posts/ what-does-the-word-security-mean

12 Voice+Code, "What Role Does the UX Team Play in Security? With Michael Snell," *Human-Centered Security* (podcast), July 20, 2022, https://share.transistor.fm/s/662675e1

13 Michelle Finneran Dennedy, Jonathan Fox, and Thomas R. Finneran, *The Privacy Engineer's Manifesto: Getting from Policy to Code to QA to Value* (New York: Springer Science+Business, 2014).

TABLE 1.1   USERS MAY HAVE DIFFERENT EXPECTATIONS OF WHAT
SECURITY TERMS MEAN

Security Term	Definition	What Your Users Might Expect
Risk	Something bad that might happen.	Users' definitions of risk vary. Product managers, security teams, and legal teams also have their own perspectives on risk as it applies to your product and organization.
Information	Data given enough context to mean something. Check out the sidebar "Types of Information You Need to Protect."	Users expect you to protect certain information (even if that is not the reality) or may question why your product is asking for certain information.
Information Systems	Devices, web apps, software, databases, and all the code that runs them.	Users expect their experiences with digital services, products, and devices to be safe.
Confidentiality	Only authorized parties can access the protected information.	Users expect certain information (like healthcare information, for example) to be safeguarded and not shared without their consent.
Integrity	You can trust that the information has not been modified or manipulated. Authenticity is also tied to integrity: you have confidence you know where the information came from.	Users expect to be able to trust your product and the information it contains. For more information, check out the sidebar "There's A Lot to Unpack When It Comes to Integrity."
Availability	Authorized users can access the information and information systems when they need to.	Users expect they can access their information when they want it. They can be frustrated when they have to solve a CAPTCHA (Completely Automated Public Turing test to tell Computers and Humans Apart). Similarly, users will be frustrated if they are locked out of their account.

Users will probably never use the security term *integrity*. For them, it's about *trust*. And they tend to have many expectations around what they should be able to trust about your product. For example, users expect the following:

- Their retirement account balance to be accurate when they look it up online.

- A social media post from a reputable organization like the SEC (U.S. Securities and Exchange Commission) is legitimate and not posted by hackers who took over the account.[14]

- They don't need to be skeptical of the voice on the other end of the phone. Instead, in 2024, some voters received a call and heard President Joe Biden's voice (generated by AI) telling them to hold off on voting in the New Hampshire primary elections.[15]

- The website from a well-known organization is just as safe to visit today as it was yesterday. They won't be thinking about the possibility that threat actors have compromised the website and the website is now prompting visitors to install malware.

- Their car will stop when they press the brakes (or the car will brake automatically when it encounters an obstacle), not that a hacker has been able to disrupt the car's brake system (an example Bruce Schneier uses in *Click Here to Kill Everybody*[16]).

---

14 Hannah Lang and Suzanne McGee, "SEC Probing Fake Post on Its X Account, Bitcoin ETFs Not Yet Approved," *Reuters*, January 9, 2024, www.reuters.com/technology/us-sec-has-not-approved-bitcoin-etfs-social-media-account-compromised-2024-01-09/

15 Tamara Keith, "New Hampshire Investigates a Robocall That Was Made to Sound Like Biden," *NPR*, January 23, 2024, www.npr.org/2024/01/23/1226251688/New-Hampshire-investigates-a-robocall-that-was-made-to-sound-like-biden

16 Bruce Schneier, *Click Here to Kill Everybody: Security and Survival in a Hyper-Connected World* (New York: W.W. Norton & Company, 2018), 79.

TYPES OF INFORMATION YOU NEED TO PROTECT

A string of five numbers is *data*. But when it's used to designate a certain geographic area, it's *information*: what we in the United States refer to as a *zip code*. A zip code plus gender plus date of birth is a special kind of information: information that can be used to identify a single person— something that is protected by privacy laws. Privacy expert Latanya Sweeney, through research she conducted in 2000, discovered you can identify over 85% of people in the United States by combining these three data points. Keep in mind that not all information needs to be protected— the percentage of car owners in a zip code probably doesn't need special safeguards—but personal information does.

Certain information is protected by regulations and standards, so if your product works with any of the following (and really, most products do), you should rally folks on your security, legal, compliance, and privacy teams. These teams have expert knowledge that you'll need to leverage (more about collaborating with those teams in Chapter 4):

- Financial information
- Information that could be used to identify a person (often referred to as *personally identifiable information [PII]* in the United States or *personal information [PI]* in Europe)
- Information related to someone's health or healthcare
- Intellectual property or business information
- Information that your users expect you to protect
- Information that could be valuable in the future—information perhaps you (and certainly not your users) haven't even thought about yet. Were you thinking about needing to protect a scan of your face fifteen years ago? I certainly wasn't.

Shift Your Mindset

In this chapter, you learned:

- Much of security happens behind the scenes, where the user is unaware of it.

- When security does impact users, it's often disruptive and confusing (as Jared Spool says, the security user experience SUX). And threat actors are looking for every opportunity to take advantage of vulnerabilities. In other words, the odds aren't in your users' favor.

- It helps to visualize the dynamics of the security ecosystem: Alice is your user, Charlie is where security impacts Alice, and then, of course, there is the threat actor, who you may know very little about. All three of these players impact one another. And everything in the system impacts them.

- Security impacts the user at every part of the user journey. In fact, security-related decisions that users make early on in their user journey (like setting up multifactor authentication—or not) can impact them later in their user journey (when their account is hacked).

- As a UX team, you have a unique skillset (and a foundation called *human-centered design*) to help improve the security user experience. But you can't do it alone—you need your security UX allies.

- Security, in the context of this book, means protecting information and information systems. But this might not be how your users think about security.

- Because of this, it's worth taking the time with your cross-disciplinary team to define what security and privacy mean for your product.

- While there's no one answer to improve the security user experience, the most important thing you can do is to ask better questions.

CHAPTER 2

The Players in the Security Ecosystem

I t's 4:59 p.m. right before a holiday weekend. Alice is almost finished with her last-minute report. She's about to press Submit, let out a big sigh, grab her coat, and walk out the door. Holiday weekend ahead!

Then the phone rings. Alice considers ignoring it, but more out of habit than anything else, she answers it. Before she can say a word, she hears Charlie on the other end. "Hey Alice, it's Charlie. Do you have a minute? We've got a problem."

Alice doesn't work with Charlie all the time. But when she does, it's usually a frustrating experience.

Alice groans internally. "Hi, Charlie. What seems to be the issue?" Maybe she can pretend there is something wrong with the connection, and Charlie will go bother someone else.

Charlie replies, "We just got in a new shipment today, and we've been doing inventory. The ATP is now QRT, and we have no idea what happened. I think some stuff got stolen."

Alice puts her head in her hands. Charlie is always using words and acronyms she doesn't understand. "Charlie, I have no idea what ATP and QRT are."

"You use the QRT when the ATP is not working."

"Charlie, that makes zero sense to me. And what do you mean you 'think some stuff got stolen?' Did anyone see anything happen?"

Charlie tends to overact, and this isn't the first time Alice has heard, "I think some stuff got stolen," only to find out someone simply messed up the inventory numbers.

Alice wonders why Charlie can't explain things in words she can understand? And why today? Can't someone else deal with this? Maybe when the manager comes in next week, she can re-count the inventory? "I just want to go home," she thinks to herself.

Pay attention to the dynamic between Charlie and Alice. It's strained. Charlie isn't the thoughtful, helpful, collaborative coworker Alice wishes he was. Instead of solving problems on his own, Charlie just sets these problems on Alice's desk, often at the most inopportune times. Worse, the way Charlie explains things is unhelpful, and it only serves to exasperate Alice. Charlie uses words that Alice doesn't understand. Ultimately, Alice doesn't trust Charlie and often questions anything he says.

You can't improve the security user experience—or security outcomes—until you improve the relationship between Alice and Charlie.

Charlie Is Ineffective Because of Competing Priorities

Charlie is where security impacts your users. As described earlier, think of Charlie as the coworker who always manages to knock on your door with a problem right before you are about to leave on a Friday afternoon.

He's often overreactive—bringing problems to Alice's attention that are really just false alarms. Alice isn't sure when she can rely on Charlie.

Further, Charlie often uses words Alice doesn't understand. When he's explaining a problem, Alice finds it hard to wrap her head around what he's trying to say. Charlie frequently fails to provide a potential solution to the problem, so Alice is left to figure out and solve the mess on her own.

As the "manifestation" of the security user experience, Charlie is often a mashup of the different, competing priorities within your organization (which explains why Charlie can often be unhelpful and annoying).

In other words, your product may be encouraging behaviors that may not be in the best interests of your users. For example, you could have the most lax privacy settings enabled by default. In order for users to change them, users have to go find them, interpret what they mean, and then decide how to proceed. Or you might give users a false sense of security that your product doesn't actually offer. Your marketing site says, "We take security seriously," but you don't guide the user in choosing the most secure configuration of your product— you leave it up to the user to figure it out.

There isn't a simple solution to the problem. And this isn't something that's necessarily new for you as a designer, because you often find yourself caught in the middle of competing priorities—and security UX is no different. Here is a breakdown of some of the competing factors you may encounter (also shown in Figure 2.1):

- **Product managers** feel pressured to build new products and features quickly, get users to sign up, and keep them using the product.

- **Leadership** feels pressured to focus on profit. They need to do a cost-benefit analysis on deciding to address security risks. They may not see a return on investment in focusing on the security user experience.

- **The engineering team** often feels pressured to do the impossible: build something that's never been done before *and* build it by tomorrow.

- **The security team** understands better than anyone else at the organization the threats to confidentiality, integrity, and availability of information and information systems facing your product. They want to make life really difficult for threat actors. An unintended consequence is, often, that these actions also make life really difficult for Alice.

- **The security, compliance, and legal teams** want to be as careful as possible before releasing your product into the world, because they are worried about complying with laws and regulations and potentially facing fines or lawsuits. There are an ever-growing number of state, federal, and international security and privacy laws and regulations to be considered.

Just because people in your organization have different priorities doesn't mean those priorities are inherently wrong. They are just approaching the security user experience from different angles. That can result in a disjointed and sometimes unhelpful experience for Alice. Threat actors know this and will capitalize on it, as you'll learn later in the chapter.

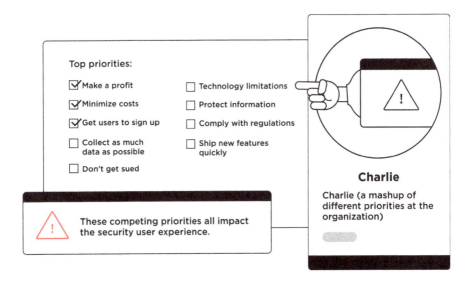

Top priorities:

☑ Make a profit   ☐ Technology limitations

☑ Minimize costs   ☐ Protect information

☑ Get users to sign up   ☐ Comply with regulations

☐ Collect as much   ☐ Ship new features
    data as possible   quickly

☐ Don't get sued

⚠ These competing priorities all impact
the security user experience.

Charlie

Charlie (a mashup of
different priorities at the
organization)

FIGURE 2.1
Charlie is where security impacts the security user experience. But he is also
a mashup of all the different priorities at your organization. Because of this,
Charlie can often be very confusing and inconsistent to Alice.

When you understand what pressures your people are facing, you
can better empathize with where they are coming from and then
work toward a solution together. In other words, even though there
are competing pressures, there is hope!

But you'll need to leverage your human-centered design skills, find
the right people, and ask the right questions. More about that in
Chapter 4, "Find the Right People, Ask the Right Questions."

NOTE IS CHARLIE GOOD OR BAD?

Is Charlie "good" or "bad"? It's complicated. To the extent that
Charlie is trying to improve security outcomes—protecting information and information systems—he can be seen as a force for
good. In many ways, he may be trying to keep Alice safer. If Charlie is skewed toward influencing Alice to do things that might not
be in her best interests to pad the corporate bottom line, well,
then, his intentions are not so great. (Also check out the sidebar
"When You're in the Position to Influence Behaviors, Question
Your Priorities.") Keep in mind that Charlie has an incredible opportunity to be a force for good—to keep people safer—but only
if he is the helpful, thoughtful coworker Alice expects him to be.

WHEN YOU'RE IN A POSITION TO INFLUENCE BEHAVIORS, QUESTION YOUR PRIORITIES

The role of a UX designer is not to change behaviors. Experimental psychologist Dr. Margaret Cunningham, chief scientist at Wethos AI, says, "I genuinely don't think we can change people's behaviors."[1]

What Dr. Cunningham means, of course, is that only people can change their own behaviors. Your UX decisions, however, can influence those behaviors.

Erika Hall, design strategist and author of *Just Enough Research* and *Conversational Design*, posted a thought-provoking message on LinkedIn: "A better question for designers and technologists to begin with than 'How might we ...?' is 'What gives us the right...to make the interventions and choices we're contemplating on behalf of others?'"[2]

Not enough teams start questions with "What gives us the right...?"

And all too often, this has to do with the competing priorities: get products shipped, ship them quickly, and make a profit while doing so. As you read the "How might we?" questions later in this book, keep Erika's quote in the back of your mind.

If Erika's quote sparks your interest, here are some additional resources:

- Center for Humane Technology, a nonprofit founded by Tristan Harris, Randima Fernando, and Aza Raskin[3]
- Humane by Design, a set of principles by Jon Yablonski[4]
- "Designers, (Re)define Success First" by Lennart Overkamp[5]

---

1 Voice+Code, "We All Have Been the 'Stupid User' at Some Point with Dr. Margaret Cunningham," *Human-Centered Security* (podcast), February 10, 2021, https://share.transistor.fm/s/0d7df5af

2 Erika Hall, LinkedIn Post, October 2023. Learn more in Erika's forthcoming book, *The Business Model Is the Grid* (Mule Books).

3 Center for Humane Technology, www.humanetech.com/

4 Humane by Design, https://humanebydesign.com

5 Lennart Overkamp, "Designers, (Re)define Success First," A List Apart, May 12, 2022, https://alistapart.com/article/redefine-success-first/

One more point. As this book is going to press, we are just beginning to understand the use cases for AI. Although AI is already helping people make decisions, it will increasingly be able to take actions on behalf of users, the implications of which can get pretty scary. And yes, threat actors are using AI, too.

People who are building products using artificial intelligence are developing frameworks and guidance that ask questions about ethics, security, and privacy. Consider integrating some of the questions and guidance in these resources into your design process:

- Salesforce's "Ethics by Design" and "Meet Salesforce's Trusted AI Principles"[6]
- Google's "Responsible AI Practices"[7]
- IBM's "AI Ethics Overview"[8]
- NIST's "Trustworthy & Responsible AI Resource Center"[9]

---

6 "Ethics by Design," Salesforce, www.salesforce.com/company/intentional-innovation/ethics-by-design/ and Kathy Baxter, "Meet Salesforce's Trusted AI Principles," *Salesforce AI Research* (blog), April 28, 2023, https://blog.salesforceairesearch.com/meet-salesforces-trusted-ai-principles/

7 Google AI, "Responsible AI Practices," Google, https://ai.google/responsibility/responsible-ai-practices/

8 IBM Design for AI, "AI Ethics Overview," IBM, www.ibm.com/design/ai/ethics/

9 "Trustworthy & Responsible AI Resource Center: Knowledge Base," National Institute of Standards and Technology (NIST), https://airc.nist.gov/AI_RMF_Knowledge_Base

Understand Your User, Alice

When you encounter Alice throughout this book, think about one of your personas. When Alice encounters Charlie, what does she expect? How might she react? What might she do? And how would that action or inaction impact Alice's security outcomes?

I gave Alice a name when I realized that I had started to detach myself from the people I was designing for—a tough thing for someone like me in UX to admit. When I gave security advice, it seemed so hollow. Some of my advice was hugely impractical. I needed to understand Alice better.

The amount of conflicting security advice that Alice reads is staggering:

"Do this, don't do this."

"You've been told to worry about this, but this isn't actually a proven threat."

"You saw this in your corporate awareness training. They told you never to do this thing. But ignore it."

NOTE HOW DO I KNOW THIS? PERSONAL EXPERIENCE

I've been researching the security user experience for years. I read lots of security advice. I've read lots of security research. I still get confused about what "good" advice is.

It's no wonder that Alice finds Charlie annoying. He can't tell her one thing and stick with it! Further, if Alice doesn't trust Charlie, every time Charlie pops his head in the door—every time Charlie impacts the user experience—she's likely to tune him out. This is known as *habituation*, by the way. It is the reason why you have to think hard about the color of your house door. You've seen it so many times that it's no longer something you pay attention to.

Additionally, some security advice (for example, when you use a password manager or two-factor authentication) can be extremely difficult for someone with a disability. Of course, you care about your users. But really let it sink in what you are asking your users to do when it comes to security. Check out the section "A Diverse Team Makes Security Usable and Accessible" in Chapter 4, "Learn and Iterate." and "Include Users with Disabilities in Your Research" in Chapter 7.

Remember that Alice is constantly multitasking. She is signing her kid up for soccer camp, trying to get her taxes done on time, scheduling doctor's appointments, reserving a spot for their camper at the lake during the summer, trying to keep an eye on her kid's phone activities, and on, and on, and on. More often than not, she needs technology to do these things.

As a UX designer, it should come as no surprise to you that, as Sasse et al.,[10] Whitten and Tygar[11] and other usable security researchers have pointed out in these foundational articles and research, security almost always gets in the way of the user's primary task. And, as you've likely experienced yourself, the interruption is typically not welcome (Figure 2.2). In other words, Alice finds Charlie annoying.

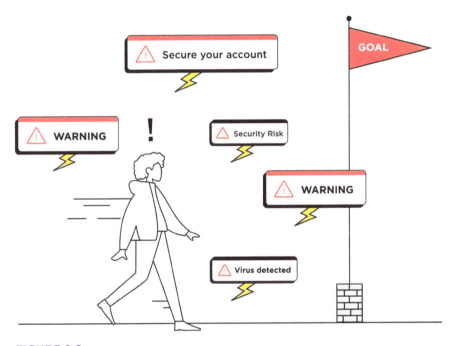

FIGURE 2.2
Alice finds Charlie annoying. He often gets in the way of what Alice is trying to do.

10 M. Angela Sasse, Michelle Steves, Kat Krol, and Dana Chisnell, "The Great Authentication Fatigue—and How to Overcome It," in *Cross-Cultural Design, CCD 2014, Lecture Notes in Computer Science,* vol. 8528, ed. P. L. P. Rau (Cham, Switzerland: Springer, 2014), https://doi.org/10.1007/978-3-319-07308-8_23

11 Alma Whitten and J. D. Tygar, "Why Johnny Can't Encrypt: A Usability Evaluation of PGP 5.0," *SSYM'99: Proceedings of the 8th Conference on USENIX Security Symposium* (August, 1999), 14, www.usenix.org/publications/library/proceedings/sec99/whitten.html

In fact, usable security researchers Angela Sasse, Adam Beautement, and Mike Wonham researched the cost-benefit analysis that employees go through as they face what they perceive as security roadblocks throughout their workday. They found that employees have a limited "compliance budget."[12] After a certain point (the 90th time you have to enter your password, the 20th time you have to log in to the VPN—virtual private network), users only have so much patience and mental and emotional energy before they throw up their hands and think, "I know I'm *supposed* to do it this way...but I just need to get this thing done!"

Psychology professor James Reason refers to these workarounds as intentional violations. Alice knows she shouldn't do something, but she does it anyway.[13]

> **NOTE YES, EVEN SECURITY EXPERTS ARE HUMAN**
>
> I've had security experts explain they fell for (or almost fell for) a scam. Or they made a configuration mistake that put their organization at risk. I've even had experts tell me they don't always practice what they preach. We are all fallible, and we don't always make the best decisions, despite our awareness of the risks.

Charlie Sometimes Fails to Help Alice

According to the "2023 Verizon Data Breach Investigations Report," which analyzes security incidents at organizations, "74% of all breaches include the human element, with people being involved either via error, privilege misuse, use of stolen credentials, or social engineering."[14]

---

12 Adam Beautement, M. Angela Sasse, and Mike Wonham, "The Compliance Budget: Managing Security Behaviour in Organisations," *Proceedings of the 2008 New Security Paradigms Workshop (NSPW '08)*, (August 2009), 47–58, https://doi.org/10.1145/1595676.1595684

13 James Reason, *Human Error* (Cambridge: Cambridge University Press, 1990), 195–197.

14 "2023 Data Breach Investigations Report: Summary of Findings," Verizon, 2023, www.verizon.com/business/resources/reports/dbir/2023/summary-of-findings/

Could some of these incidents have been prevented if Charlie helped Alice? For one, Charlie might be able to help prevent what psychology professor James Reason refers to as mistakes (further broken down into knowledge-based mistakes and rule-based mistakes), slips, and lapses.[15]

Knowledge-Based Mistakes

Alice may not set up an account or a product in the most secure manner possible because she doesn't know how—or may not even be aware that the security settings exist. I once spoke to someone for twenty minutes about two-factor authentication before she admitted that she had no idea what two-factor authentication was. This person would never, ever go looking for how to set up two-factor authentication—she doesn't know it exists! With increasingly complex systems, even security experts don't fully understand how every system works.

Rule-Based Mistakes

There are a lot of what James Reason might refer to as "rules" related to security. Some rules are downright challenging to carry out. And, unfortunately, due to the nature of the complex, ever-changing security ecosystem, many rules don't always withstand the stress test of real life.

Remember, threat actors know these rules and "best practices," too. And they are looking for ways around them.

For example, Alice may know to check an email's display name (i.e., Bob Smith) against the "from" address (a scammer email address that might look similar to Bob's email address). Alice may even know how to check if the email was authenticated.

Unfortunately for Alice, performing these checks won't help her if Bob's email account has been taken over by a threat actor. It also won't help her if a threat actor is using a legitimate service (say, an online payment company or invoicing system) to send fake invoices or requests for payment.

---

15 Reason, *Human Error.*

In both of these cases, Alice can't rely on the rules she's been taught to determine if an email is spoofed or not. Ideally, Alice will need to use another method for verifying the contents of the email (meaning Alice now needs to stop whatever she was doing and start an entirely different task altogether). Or she might try to read the email's contents and look for clues there. The contents of the email, however, will be cleverly crafted to play on Alice's emotions. Threat actors don't want Alice to *think*—they want her to *act* now. The email may encourage Alice to call a (scammer-operated) 1-800 number or click on a malicious link.

The takeaway? Even though Alice might know the security-related "rules" in a given scenario, threat actors look for ways to get around these rules or render them ineffective. And threat actors will only get better and leverage different types of communication channels: phone calls, text messages, social media messages, and more. It's up to us—people designing products—to help Alice stand a chance at protecting herself.

NOTE PERSONAL SECURITY TIP

> If you receive a message (via carrier pigeon, telegram, phone call, text message, email, social media, chat, or anything) that needs you to do something fast, seems too good to be true, or seems to pull at your heartstrings, the best thing you can do is slow yourself down. Look for an alternative way to verify the information. Close the message or put down the phone and call the friend or family member directly. Visit the bank website using a bookmark and check your notifications there. The organization's website will have the official phone number.

Slips

Alice might inadvertently click on a phishing link while on her phone at her kid's soccer game, not because she didn't realize the message was a phishing message, but because she was juggling multiple things and her thumb hit the link. Alice might also share a spreadsheet with employee salaries to John Smith, the external contractor when she meant to share it with John Smith, her colleague in HR. Sometimes, Alice realizes the slip quickly—sometimes, she never does.

Lapses

Alice may forget to change default credentials or miss an important step in configuring an account securely. She *knows* she has to do it, but she simply forgets. You can imagine that the greater the number of steps and the more pressure Alice is under, the more likely she is to have a lapse.

Remember, threat actors know all of this already—often better than the people designing the product. They are ready to take advantage of common mistakes, slips, and lapses.

Threat Actors Take Advantage of Any Vulnerability

There are always threat actors lurking in the background, and Alice knows very little about them.

They take advantage of what Alice does or does not do. For example, she might not regularly update her software, or she might not set up a device, service, or account in the most secure way. Threat actors also might try to trick Alice into doing something (like give up her account credentials or passcode) or manipulate her (make her believe something is true). Or a combination thereof!

Threat actors prey on the fact that Alice is human.

There are underlying patterns in how threat actors get what they want. No matter what you are working on, no matter what new technology is developed, these patterns will continue. This particular list focuses on how a threat actor takes advantage of Alice—things you need to pay particular attention to—and anticipate—as part of designing the user experience. They are posed as questions that a threat actor might think to themselves, summarized from MITRE ATT&CK®,[16] a resource that documents threat actor tactics and techniques: *How do I trick Alice into giving me money, assets, access to the information or system I want access to, or believe something I want her to believe?*

---

16 MITRE ATT&CK framework (attack.mitre.org) is a publicly-available resource of threat actor tactics and techniques (and methods of combating them). It's updated regularly and can give you a sense of the many ways threat actors exploit both human and technical vulnerabilities. MITRE ATT&CK wants to ensure that I'm clear I'm not implying I'm affiliated or endorsed by it—I'm not.

For example:

- **How can I convince Alice I am someone or something Alice knows and trusts?**
  - A person?
  - A business or brand?
  - A website Alice regularly logs into?
  - A dialog box or notification Alice is used to seeing? For example, a fake operating system dialog box that asks Alice to input her username and password.
- **How might I scare, shame, or trick Alice into doing or giving me what I want? Where can I capitalize on moments that are particularly stressful or emotional for Alice?**
  - A job opportunity during a down job market?
  - An update during a pandemic?
  - A threat to shut off electricity for an unpaid bill?
  - A loved one who is in trouble and needs money?

Unfortunately, this list goes on and on: *Where has Alice left a potential entry point so I can get access to the system?*

For example:

- **Where has Alice likely set up the product insecurely?** Alice may forget to change default credentials, write insecure code, or miss a step in securely setting up an account or system.
- **How can I leverage automated tools to scan for potential entry points?** Threat actors might use passwords obtained from a hacked database from Product A and try them out on Products B, C, and D. This is known as a *credential stuffing attack*. Or they might scan for unpatched software and scan for potential entry points (by trying default credentials that were never changed, for example) using automated tools.

Here are a few additional points that are important to note:

- **Charlie can inadvertently help threat actors.** When Charlie (remember, Charlie is where security impacts the user experience) makes it hard for Alice to understand the system and set it up in the most secure way, entices Alice to create workarounds, or provides confusing or inconsistent messages, Charlie is helping the threat actors. Threat actors know where these things are most likely to happen. And they capitalize on these moments.

- **Threat actors constantly adapt.** And, unless you address the security user experience through a human-centered design process, their tactics will only improve. As of the writing of this book, organizations are racing to adopt AI (artificial intelligence), shove it into all of their existing products, and leverage it to build new ones. AI will only make it easier for threat actors to do things like send very convincing, targeted phishing emails, as well as new and novel things we haven't even anticipated yet.

- **Not all employee intentions are good** (also known as an *insider threat*). Employees and other people who have access to the system (such as vendors) might also intentionally or maliciously try to steal information or disrupt or damage systems. This is particularly problematic when these people have access to information or systems that most people don't have—such as the ability to access sensitive customer information or the ability to steal or alter source code.

How might a threat actor leverage your product or platform for their own motivations? How might the way Alice understands, uses, sets up, configures, or updates your product make it easier for threat actors to get what they want? How might Charlie—where security impacts Alice—help her stay safer? You need to understand the ecosystem (Figure 2.3).

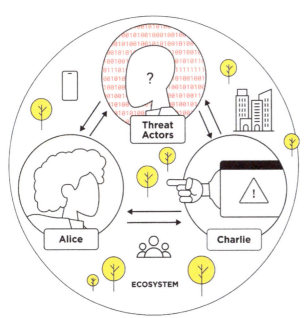

FIGURE 2.3
When you're designing the security user experience, you need to account for the dynamics between Charlie, Alice, and threat actors. The actions of one influence the actions of the others.

Shift Your Mindset

In this chapter, you learned:

- There are three players in the security ecosystem (Charlie, Alice, and the threat actor), and they all impact one another.
- *Charlie* is where the security user experience impacts Alice.
- *Alice* is your user. She often finds Charlie annoying and unhelpful. That means she is less likely to trust him.
- The *threat actor* understands the dynamics between Charlie and Alice, and capitalizes on that tense relationship. Further, the threat actor looks for any opportunity to take advantage of Alice.
- It's critical for you and your team to understand the dynamics between these three players in order to improve the security user experience.

Beware of Unintended Consequences

My coffee grinder, shown in Figure 3.1, won't work unless the cover is completely on in a locked position—where it would be impossible for my finger to be near the blade. This prevents me from making a dumb pre-caffeine, finger-losing mistake in the early morning.

Now, grinding coffee beans is a means to an end. I want a quality cup of coffee and grinding coffee beans is one step on the way to getting there. Cutting my finger using the grinder is honestly one of the last things on my mind. Yet the risk is still there.

I could get distracted, fail to read the safety instructions that accompanied the coffee grinder (Ha! Who does that?), or my young niece could get hold of it. I, my niece, or any other members of the household introduce risk just by being human.

FIGURE 3.1
In order for the blade to activate, you must slightly twist the coffee grinder lid, locking it into place, making it impossible for your finger to be near the blade while it's in motion. The coffee grinder manufacturers intentionally made it slightly more difficult to close the lid in an effort to make the coffee grinder safer.

The coffee grinder manufacturers know how careless people can be, so they intentionally make the cover slightly more difficult to put on. That's right, the coffee grinder designers made my life a tiny bit harder on purpose. They introduced friction into the user experience.

That's because the coffee grinder manufacturer is *managing risk*. Sure, they don't want me to chop off my finger. But they also don't want to deal with the bad press and lawsuits that result from a customer's severed digits.

Your security colleagues are also trying to manage risks—security risks. That's because you, me, and every well-intentioned person we know introduces risk just by way of being human. (Remember Alice, your user? Yeah, she's human.) Humans are fallible. They make mistakes, and they don't always do what they are supposed to do. And threat actors are always trying to take advantage of, trick, and manipulate Alice.

As you read in Chapter 2, "The Players in the Security Ecosystem," people at your organization—product managers, engineers, leadership—also have their own priorities and risks they are trying to manage (things like limited engineering resources, customer churn, and time-to-market).

Managing risk—security or otherwise—often means making trade-offs, just like the coffee grinder manufacturers had to make. As you'll learn in this chapter, you need to be careful that the consequences of those trade-offs don't take you by surprise.

The Dynamics of the Three Players Can Surprise You

The evolution of multifactor authentication (MFA), also commonly referred to as *two-factor authentication* (2FA), is an example of how the actions of Alice, Charlie, and threat actors influence one another. 2FA is when two factors are required to sign in. MFA means two *or more* factors are required to access a service.[1] The "factors" in MFA or 2FA could be something Alice knows, something Alice has, or something Alice is.

---

1   It's confusing enough having 2FA and MFA used interchangeably (and often referring to exactly the same thing). But, just to make things more confusing for Alice, there's another acronym she might see: 2SV. Two-step verification (2SV) means that Alice might use two of the same authentication factors. So maybe she uses both a password and enters in her mother's maiden name. These are both things Alice knows. 2SV is considered less secure than MFA and 2FA.

If Alice has MFA set up on her bank account, for example, she'll be asked for both her password (something she knows), as well as be prompted to enter a time-sensitive code sent by text message or generated by an authenticator application (something she has). The greater the number of factors, the greater the security will be. But, as you may have experienced yourself, it often comes at the cost of usability.

By the way, these factors will evolve over time.

The rest of this chapter is a very simplified story of how multifactor authentication came to be and evolved. Note that it accounts for only a couple out of a myriad of factors that impacted the adoption and evolution of MFA. (This is not meant to be a scholarly article on the evolution of MFA.)

The most important takeaway: the actions of Alice, Charlie, and the threat actors impact the security user experience.

In this scenario, imagine that Alice is accessing her work email account.

Passwords Just Won't Cut It—Enter Multifactor Authentication

Before MFA, Alice entered her username and password to access her work email account.

Charlie: "Alice, I need you to create a unique password for this account. The password needs to be at least 8 characters, include at least one uppercase and one lowercase letter, as well as a special character."[2]

Alice: "Charlie, I need to access my work email so I can do my job. I can't possibly remember all the passwords for my work accounts, so I'll just choose one and use them on every account."

The threat actor thinks to themself: "I know how the Alices of the world operate—they use the same passwords on all of their accounts. I'll blast out some phishing emails and see who falls for them. Or I'll just take a password I bought on the dark web that works for Product A and see if it works on her email account."

Eventually, one of Alice's accounts is compromised.

---

2   Christian Rohrer talks about some of the reasons that passwords are problematic in: Voice+Code, "How to Design Great User Experiences in a Complicated Cybersecurity Ecosystem with Christian Rohrer," *Human-Centered Security* (podcast), January 6, 2021, https://share.transistor.fm/s/a88f7010

Alice's employer decides to force Alice to use multifactor authentication. The intention? It's not to be mean or difficult. It's there to manage security risk. Alice's employer is trying to do the same thing that the coffee grinder manufacturers did.

Alice's employer decides they'll give Alice the option to respond to a push notification sent to one of her devices. All she has to do is tap "yes" when she sees the notification on her phone or smartwatch, as shown in Figure 3.2.

Push Notifications Are Easy for Users *and* for Threat Actors to Exploit

What Alice's employer didn't account for, though, is how much Alice hates multifactor authentication. Not dislikes it. She *hates* it.

Charlie, "Now that you've entered your password, Alice, wait a moment while I send a push notification to your phone. You'll need to choose 'yes' to finish signing into your account."

Alice, "Ugh, Charlie! My phone is in the other room!" Alice scrambles to find her phone, "Wait, where did the notifi-

FIGURE 3.2
After Alice enters her password to access her email account, she may receive a prompt on her phone asking her to confirm that she is signing in. This is one implementation of multifactor authentication—one that threat actors like to take advantage of.

cation go? Did I miss it?" Meanwhile, Alice gets distracted responding to text messages and looking at a myriad of social media notifications.

NOTE CONSIDER WHAT YOU ARE ASKING OF USERS

Amy Grace Wells, associate director of content design at 10up and who has ADHD, encourages designers to consider the steps that users have to go through to perform a task like authenticating with MFA. For example, for a six-digit MFA code, a user has to find their device, unlock it, open the right app (Is it sent via text message? Or one of several installed authenticator apps?), remember the code, and enter the code on a separate device. All under time pressure (many codes generated by authenticator apps, for example, expire every 30 seconds). And then there is the added distraction of unlocking your phone and feeling compelled to look at notifications or open a social media app.[3]

The threat actor thinks, "It's still easy enough to get Alice's password. Now how do I get around this MFA business? What if I just bombard Alice with push notifications? I bet she'll get so annoyed or confused that she'll say 'yes' to one of them."

The threat actor puts the plan into action.

Some time goes by. Alice is at her kid's soccer game and is half-paying attention as she answers some work emails. Suddenly, she gets a bunch of push notifications all at once, all from her authenticator app. She thinks, "Wait, what are these from? Maybe this is my email app on my laptop trying to sign in? Is this because I just checked my work email on my phone? Wait, did I just tap on something?"

Unfortunately, it's the threat actor who is signing in as Alice, triggering the push notifications.

This threat actor tactic even has a name: *MFA fatigue*. MFA fatigue is when users are overwhelmed or confused by push notifications being sent. This is made worse by the fact that Alice often has many different services all sending her push notifications for just about everything.

To combat MFA fatigue, most systems will now lock the account if someone tries to log in multiple times without successfully completing multifactor authentication. Because of this policy, threat actors need to be selective about how many times they trigger MFA.

Of course, this does not stop threat actors. Instead, the threat actors came up with a variation of the tactic that has proven to be effective: the threat actor calls, texts, or emails Alice posing as someone from her organization (her IT team, for example). The threat actor tells Alice she needs to say "yes" (or accept) the push notification because they are working on transferring her to a new system, or because she's been locked out of her account, or they need to verify her identity. Tricky. But this is exactly what Alice is up against.

---

3   Based on a conversation with Amy Wells on November 30, 2023.

Enter the Modified Push Notification—with a Less-Usable Twist

One of the ways that organizations can combat MFA fatigue is by making users like Alice work a little harder. So instead of just saying "yes" to a push notification, Alice has to enter a code generated by the service requesting access. If the user doesn't have the code, they have no other option than to choose "No, it's not me."

Is it a loss for usability? Well, yes, it is.

Of course, the intention behind mechanisms like MFA is not to create bad user experiences. It's trying to manage risk. But you can see the dynamics at play here. It's like watching a three-way ping-pong match. The employer introduces two-factor authentication because the organization is trying to manage risk. The user reacts. Then the threat actor adapts. The organization tightens down security even more. Then the user reacts. The threat actor adapts again. You get the idea. It never ends. And it will never end. That's why you need to iterate (learn more in Chapter 7, "Learn and Iterate").

And there's one other important thing you can do: help your security UX allies understand Alice better.

Help Your Team Understand Alice Better

Your security UX allies (see Chapter 4, "Find the Right People, Ask the Right Questions") can help you anticipate the security threats facing your system.

But you understand Alice. Help your security UX allies channel their inner Alice. Blair Shen, a product designer at Duo, explains that you need to think from different perspectives: the risk the user introduces, and the risks threat actors introduce, all while balancing business needs. She says: "We work with our security engineering team so they can help us think like a hacker and predict some of the potential vulnerabilities in any possible design solutions we come up with. As a design team, we see security from the user's perspective."[4]

---

4   Voice+Code, "Designing Multi-Factor Authentication with Blair Shen and Bethany Sonefeld," *Human-Centered Security* (podcast), October 19, 2022, https://share.transistor.fm/s/52f79a3b

Every design decision you make has consequences you may not anticipate. Users—and threat actors—will do things you didn't account for.

What are the unintended consequences of this design decision? What might users do or not do that would impact security outcomes? What might threat actors do that would negatively impact security outcomes? These are tough questions to answer.

Here's your challenge: As you'll learn in the next chapter, you need to find the right people and ask the right questions to find the right solution for your organization and circumstances.

Shift Your Mindset

In this chapter, you learned:

- Your organization is trying to manage security risks, but there is a dynamic at play. Any time Charlie pops up, Alice reacts, and the threat actor reacts. Charlie adapts or evolves to account for threat actor actions, then Alice reacts again, and ultimately the threat actor adapts.

- This dynamic is really challenging to design for! There will often be unintended consequences for any security user experience decision.

- Know that you will need to be open to iteration. It needs to be part of your design process. But also know that the biggest contribution you can make to improve the security user experience is by helping your security UX allies understand Alice.

Find the Right People, Ask the Right Questions

As a designer, you have long practiced putting yourself in the mindset of your users. In fact, you've probably said things along the lines of, "Will the user really know how to do this? Frankly, *I'm* confused, and I'm on the team that designed this."

These are exactly the types of questions you need to ask when it comes to improving the security user experience:

"Will users understand this warning? I have a feeling they won't know what these words mean."

"Won't users just ignore this warning and figure out another way to do it? And won't that make the problem worse?"

You are an expert on your users, which means you can anticipate where they might be tricked or manipulated by threat actors. If you're working on a peer-to-peer payment app, don't be afraid to ask, "Could users be tricked into sending money to a threat actor? Would they have any recourse for getting that money back?"[1]

These aren't dumb questions. Obviously, someone getting tricked into sending their money to a threat actor would be a terrible user experience.

The one thing you shouldn't do is stay silent. And you shouldn't wait until the eleventh hour to have these conversations. At that point, it could be challenging to make meaningful changes, and as everyone knows, when you push things to the next meeting, they get moved to the next sprint, then to the next quarter, and then they just never happen.

Just make sure that you aren't asking these questions in the comfort of your UX-only vacuum. You have to involve a cross-disciplinary team.

Find Your Security UX Allies

What is one universal piece of advice that fellow UX designers have for improving the security user experience? The answer: Have conversations with your cross-disciplinary teammates early and often. These cross-disciplinary folks are your "security UX allies." I use this name because the people you include to improve the security user

---

1   Some of your cross-disciplinary team members might refer to these as "abuse cases." Check out OWASP, "Abuse Case Cheat Sheet," https://cheatsheetseries .owasp.org/cheatsheets/Abuse_Case_Cheat_Sheet.html

experience are not just stakeholders, they are allies, even if it might not seem so at first.

Kristen Seversky, a former developer who is now in a product operations role, has, over the course of her career, seen major projects get derailed due to failure to involve the security, legal, and compliance teams. Looking back, she realized that if her team had known what questions to ask early on, it would have saved a lot of time and frustration.

Kristen's advice? Every time you start a project, you should plan ahead to identify and work with teams such as legal, compliance, engineering, and security, and build in this collaboration process as part of your overall project timeline.

Who Should Be Involved?

You may have to do some organizational digging to find folks you have never worked with before, including the following (see Figure 4.1):

Stakeholders Who Are Security- and Privacy-Focused

- Product security and application security teams (AppSec)
- DevOps or DevSecOps team
- Information security team (which may include the Chief Information Security Office [CISO] or Chief Information Officer [CIO] and the people who report to them)
- Chief Privacy Officer (CPO)
- Engineering/development team
- Quality assurance (QA) team
- Information technology team
- In-house or external legal team(s)
- Risk management team
- Compliance team

Stakeholders Who Have Direct Relationships with Users/Customers

- Product managers
- Customer experience (CX) team
- Product support
- Relationship/account managers

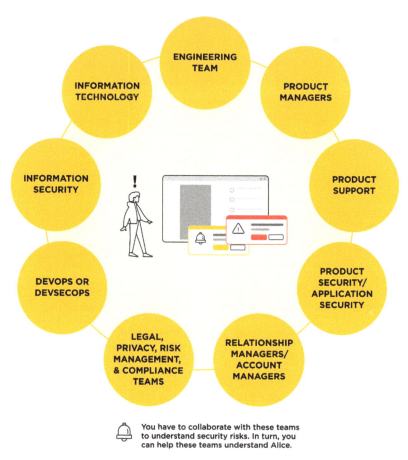

You have to collaborate with these teams to understand security risks. In turn, you can help these teams understand Alice.

FIGURE 4.1

Find your security UX allies. Together, you need to collaborate and collectively understand and account for the security ecosystem, including Alice, Charlie, and the threat actor.

NOTE JUST FIND THE PERSON RESPONSIBLE FOR SECURITY

The most important thing is to start a conversation with someone who is responsible for security at your organization—no matter what their title is.

Here's what you might say, "My team works on peer-to-peer payments, and we are designing a feature that allows users to buy and sell cryptocurrency. We want to talk though some security concerns and potential abuse cases for this feature. Who are the right people to have involved in this conversation?"

Whatever you say, make sure that you emphasize your team's desire to protect both your users and your organization.

A Diverse Team Makes Security Usable and Accessible

In order to make security usable and accessible, you need a diverse team, including people with disabilities. In addition, digital accessibility expert and former head of accessibility for McDonald's and VMware, Sheri Byrne-Haber says, "By making your organization more psychologically safe and more diverse, you're inherently going to improve your security for your customers with disabilities." Team members need to feel safe to share their own disabilities, as well as express their concerns when a security-related solution won't work for people with disabilities. Otherwise, you may find out too late—after a customer files a complaint.

Sheri has several additional recommendations for people on the UX team:

- **Give your personas disabilities.** This helps put your team in the mindset of regularly questioning what and how your persona would think, feel, need, and act in a given situation. These personas should be based on research.

- **Work with your colleagues in security and privacy** (all the more reason to encourage cross-disciplinary collaboration). Accessibility, like security and privacy, has compliance implications. When you join forces, Sheri says, "You can bolster each other's initiatives."

- **Ensure that your design system and component library are built with accessibility in mind.**

Check out additional recommendations in Chapter 7, "Learn and Iterate."

Remember, your security UX allies all have their unique superpowers. Combined, you are in a better position to improve the security user experience. You are in a better position to keep people—and your organization—safe.

So, together, how might you do that?

The Magic Question Is "What Can Go Wrong?"

The questions at the beginning of the chapter—questions that surface how users could introduce information security risks and how threat actors might manipulate or trick users—are exactly what threat modeling is all about.

While *threat modeling* might seem like a fancy term, you and your team ask these questions all the time. In other words, you are threat modelers, and you didn't even know it.

According to Adam Shostack, author of *Threat Modeling: Designing for Security*, threat modeling is about asking four questions:[2]

1. What are we working on?
2. What can go wrong?
3. What are we going to do about it?
4. Did we do a good job?

When you think about the relationship and dynamics between Alice, Charlie, and the threat actor, ask, "What can go wrong?" These four words are pretty powerful!

Adam explains the ultimate goal of asking these questions is to have more productive conversations with your cross-disciplinary teams. He stresses that threat modeling is part of an iterative process.

The questions are even more powerful if you have a lens or framework to work from. Adam uses STRIDE[3] in *Threat Modeling: Designing*

---

2   If threat modeling sounds interesting to you, I highly recommend checking out:

- Adam Shostack, *Threat Modeling: Designing for Security* (Indianapolis, Indiana: John Wiley & Sons, Inc., 2014).
- Adam Shostack, *Threats: What Every Engineer Should Learn from Star Wars* (Hoboken: John Wiley & Sons, Inc., 2023).
- Voice+Code, "Threat Modeling for UX Designers with Adam Shostack," *Human-Centered Security* (podcast), November, 9, 2022, https://share.transistor.fm/s/ad97b9b4

3   Shostack uses the mnemonic STRIDE as a framework in *Threat Modeling: Designing for Security*. STRIDE, developed by Praerit Garg and Loren Kohnfelder while at Microsoft, makes it easier to brainstorm potential threats to your system. STRIDE stands for Spoofing, Tampering, Repudiation, Information Disclosure, Denial of Service, and Elevation of Privilege. Understanding common information security threats and going through threat modeling exercises with your engineering and security team can help you better visualize threats and, as a result, ways where the user experience might be impacted.

for Security, but your UX team may want to develop your own customized framework or set of questions. One place to look for inspiration is Lorrie Cranor's "A Framework for Reasoning About the Human in the Loop," shown in Figure 4.2.[4] Think of Lorrie's framework like a user flow. In her model, the "communication" is at the beginning, which Lorrie says can be thought of broadly as, "warnings, notices, status indicators, training, policies."

At the other end is the "behavior." This is the action (or lack of action) that the user takes. Your job is to ask, "What can go wrong between the communication and the behavior?"

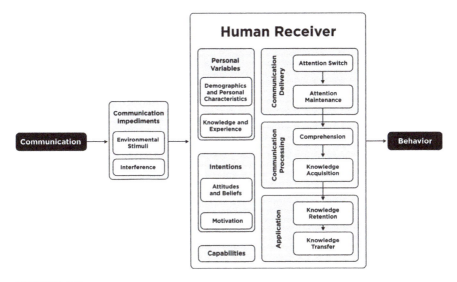

FIGURE 4.2

Cranor's human-in-the-loop security framework starts with the input or "communication" and ends with the "behavior."

Remember, channel your inner Alice. Jumping off Lorrie's framework, in a given point in the user journey, you should ask these questions:

- What might Alice expect?
- How might Alice react?
- How might a threat actor trick or manipulate Alice?

4 Lorrie Faith Cranor, "A Framework for Reasoning About the Human in the Loop," *UPSEC'08: Proceedings of the 1st Conference on Usability, Psychology, and Security* (April 2008): 1–15, https://doi.org/10.1184/R1/6620651.v1

- What are the potential unintended consequences of our design choices? (This last question is particularly important. As you learned in Chapters 2, "The Players in the Security Ecosystem," and 3, "Beware of Unintended Consequences," Charlie may cause Alice to react in ways that you did not anticipate.)

NOTE CHECK OUT THE QUESTIONS AT THE END OF "A FRAMEWORK FOR REASONING ABOUT THE HUMAN IN THE LOOP"

What you might find particularly useful about Lorrie's work are the set of questions at the end of the paper. These are questions that you and your team might incorporate into your own framework. Here are just a few examples:

- What type of communication is it? Is it active or passive? Is this the best type of communication for this situation?
- What relevant knowledge or experience do users have?
- Do users understand what the communication means?
- Does the behavior result in the completion of the desired action?

Threat Modeling in Action

Let's try something out for fun. Then I'll give some words of advice.

You're on a team responsible for designing the account and settings pages for an email service (think, Google Workspace or Microsoft 365). You've become aware of a problem that is affecting more than a few users: they've lost access to their accounts and can't get back in.

You look through customer forums and some support transcripts, and you suspect that one of the reasons is confusion around how users know their account has been compromised and the steps they need to take to regain control of it.

You channel your core persona, Alice. What would Alice do if she found out her account was compromised?

Alice would say she would change her password (Figure 4.3).

But you know it's not quite that easy. Alice is only partially right. Yes, Alice has to change her password. But there are some other complicating factors, too.

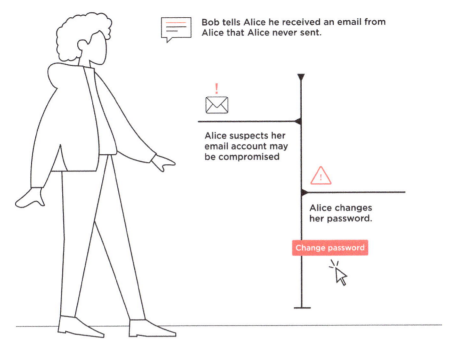

FIGURE 4.3
The first thing Alice would do if her account was compromised is change her password. This is partially right. But there are many other steps Alice needs to go through to ensure that she regains full control of her account.

Did the threat actor who has access to Alice's account change the account password, locking Alice out? The threat actor may have also changed (or set up) multifactor authentication, added or changed Alice's recovery email address, and set up mail forwarding.

Suddenly, this problem becomes quite a bit more complicated.

Worse, the complete set of steps aren't included in an easy-to-find way in your product's documentation, meaning that Alice would need to go around searching for them, compiling information from multiple sources, understanding and implementing it. Not an easy task. Here's what Alice would need to do (also shown in Figure 4.4):

1. **Change her password** (if she can't log into the account, she would first need to go through the account recovery process. For example, sign into her recovery email, click on a password recovery link sent via email, and then choose a new password for the compromised account).

2. **Check her password recovery methods.** The threat actor may have changed them or added their own.

3. **Check login and multifactor authentication methods.** The threat actor may have changed them or added their own.

4. **If there are recovery codes, regenerate them** (which inactivates the old ones the threat actor may have stolen).

5. **Revoke access to any devices that may have access to the account.**

6. **Check which third-party apps may have access to the account.** Revoke the ones that shouldn't or don't need to have access.

7. **Check if mail is being forwarded.** Even after Alice changes her password, if mail is being forwarded, the threat actor is getting every email Alice does.

8. **Check if any mail filters have been enabled.** Threat actors sometimes apply filters to automatically filter out messages that have words like "security" in them, so Alice may miss important messages.

9. **Scan her device(s) for malware.**

10. **Change passwords on accounts associated with the compromised email account.**

11. **Contact the people Alice has emailed recently or people in Alice's contact list to notify them that her account was compromised and that they may have received malicious emails from the address.**

This process requires the user to go to many different screens to ensure that their account is safe. So not only does Alice need to know all the steps to take, but she also needs to know where to go to complete those steps.

By taking the time to visualize this process, you can start to get an idea where there might be user experience challenges. Alice shouldn't have to go through so many confusing, time-consuming steps to protect one of her most important accounts. Help Alice!

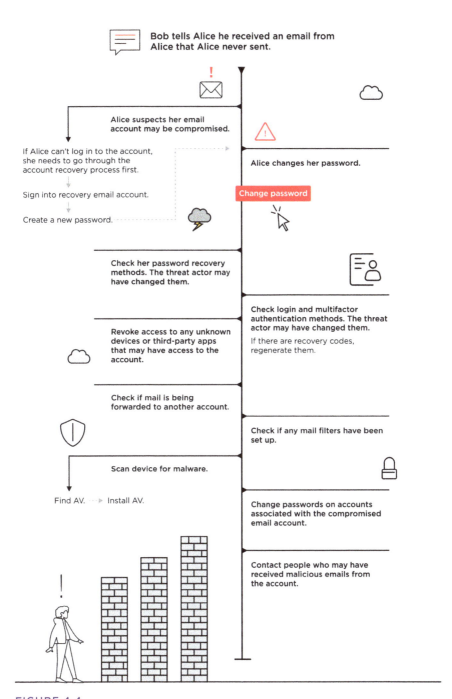

Bob tells Alice he received an email from Alice that Alice never sent.

Alice suspects her email account may be compromised.

If Alice can't log in to the account, she needs to go through the account recovery process first.

Sign into recovery email account.

Create a new password.

Alice changes her password.

Change password

Check her password recovery methods. The threat actor may have changed them.

Check login and multifactor authentication methods. The threat actor may have changed them.

If there are recovery codes, regenerate them.

Revoke access to any unknown devices or third-party apps that may have access to the account.

Check if mail is being forwarded to another account.

Check if any mail filters have been set up.

Scan device for malware.

Find AV.   ·▷ Install AV.

Change passwords on accounts associated with the compromised email account.

Contact people who may have received malicious emails from the account.

FIGURE 4.4
Alice needs to go through a series of steps to ensure her account is safe. There's no straightforward way to do this. She needs to go to different parts of the product to complete the steps.

Helping your security UX allies understand the broader context is much more effective when everyone can visualize and respond to what you're talking about, like Figure 4.4. Remember that outside teams may have very little context about what you are working on. A visual—a user flow, a journey map, a service blueprint, or screen-shots—can make a difference in getting everyone on the same page.

You might get some things wrong, or there may be some aspects of the user experience you haven't accounted for—and that's OK! The preliminary work you do to share with your cross-disciplinary team serves as a catalyst for conversation. At this point you are trying to understand the problem—not jump to solutions.

THE BEST SECURITY UX RETURN ON INVESTMENT

Here's where you might place your focus—i.e., where you'll get the best security UX return on investment:

- Anytime you are explaining security or privacy concepts—this could even be before users start using your product
- During sign-up
- When you are asking for personal or financial information
- During onboarding or configuration of your product
- When the user accesses or shares sensitive information or documents
- When users are asked to authenticate
- When the user receives a communication about privacy or security
- When the user is asked to make a security- or privacy-related decision
- When the user has to decide whom or what to trust
- When your product is connected to a device that could influence the physical world—for example, software that automatically turns on or manages the behaviors of machines or other physical devices

Michael Snell, UX Research Lead at JPMorgan Chase & Co., advises designers to start with a core set of user journeys. Then he suggests, "Break down the security implications of those user journeys."[5]

---

5   Voice+Code, "What Role Does the UX Team Play in Security? With Michael Snell," *Human-Centered Security* (podcast), July 20, 2022, https://share.transistor.fm/s/662675e1

Ask the Right Questions

Just reading the list of actions Alice needs to take probably has you reeling with questions. Yup that's you, threat modeling!

These questions all stem from Adam Shostack's threat modeling framework described earlier in the chapter. Namely: "What can go wrong?"[6]

- How would Alice know her account has been compromised? How would she know the next step to take once she has become aware?

- If we sent Alice a communication that alerted her that the account was compromised, would she question if the communication was legitimate? How might threat actors leverage these types of communications to trick Alice? How do we ensure that Alice knows she can trust the communication is coming from us and not a threat actor?

- What steps would Alice need to take if her account password and recovery options had been changed by a threat actor? What if Alice never set up her recovery options in the first place or no longer has access to those recovery options?

- How would Alice know to go through the series of steps required to regain full access to her account?

- Might Alice question the need to go through all of these steps and simply stop at changing her password? What other scenarios might prevent Alice from carrying out the steps in full and how do we account for those?

- When Alice reaches out to us for help when she's locked out, how do we help her? How do we identify that she is who she claims to be?

- If we've built in ways to recover the account, how do we prevent threat actors from using that to gain access to accounts?

- What other factors have we failed to consider (for example, different setups, the user never logs in via a web browser, etc.)?

Here's why asking these questions is important.

First, you channeled your inner Alice and anticipated where she might get bogged down in the process. In other words, you anticipated what might stand in the way of Alice successfully recovering her account and getting the threat actors out.

---

6   Shostack, *Threat Modeling: Designing for Security.*

Second, you anticipated the actions of threat actors. Threat actors might try to send Alice fake account compromise emails. Or threat actors might go through the account recovery process pretending to be Alice.

But here are a few really important things to bear in mind:

- **Make sure that everyone knows what the secure path is.** Sometimes there is a recommended way to set up your product (just like there was a recommended, yet labor intensive, way to kick the threat actors out of your email account). You just might not know what that way is. Always ask your security UX allies: What is the recommended security setup for this scenario? You have to know what the path is in order to improve it.

- **Keep the bigger picture in mind.** One of the biggest mistakes I have made is thinking about the security user experience in isolation compared to the rest of the user experience. I lost the broader context. That's not how the user experiences it—so it doesn't make sense to isolate it in terms of how you address it. To Alice, it's just her experience.

Speaking of which, don't forget about related parts of the user journey. In examining one aspect of the security user experience, you'll often realize that things the user did or didn't do at other parts of the user journey impact their security later on.

As you just read with the list of steps Alice needs to complete to recover her account, what users do and don't do during sign-up and onboarding can have a *massive impact* on their security down the road. These are places where a user could (or not!) do the following (Figure 4.5):

- Choose a strong password.
- Set up multifactor authentication.
- Associate their device with the account (such as with passkeys).
- Choose account recovery options (like linking a phone number or email address to the account or having backup multifactor authentication options).
- Configure the service or device for security (such as security and privacy settings or for roles like administrators, who are responsible for adding other users to the service).

Fully understanding related parts of the user journey may mean that you need to collaborate with other teams who are responsible for different aspects of your product or the user journey and take a step back to ensure that you are addressing the problem holistically.

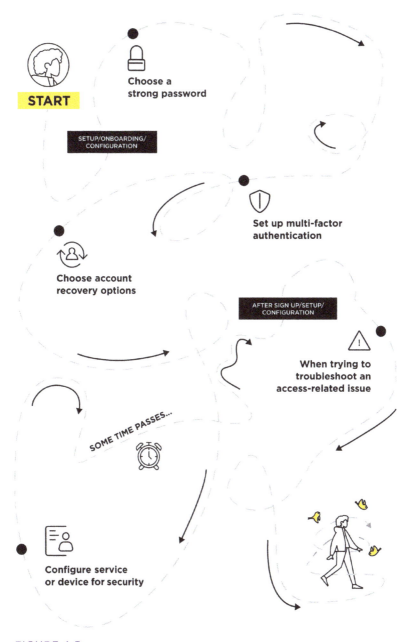

START

Choose a
strong password

SETUP/ONBOARDING/
CONFIGURATION

Set up multi-factor
authentication

Choose account
recovery options

AFTER SIGN UP/SETUP/
CONFIGURATION

When trying to
troubleshoot an
access-related issue

SOME TIME PASSES...

Configure service
or device for security

FIGURE 4.5
What users do during onboarding impacts their account security—and
security user experience—later on.

Some (or all) of your users may have the ability to make decisions or take actions that—compared to other users—can pose major information security risks to their organizations and customers. Typically, these users have privileged access (for example, they can access administrator accounts or manage employee or customer accounts, devices, and infrastructure).

If you design enterprise software, it's very likely that some of your users would be in this scenario at least some of the time. Some examples include when users need to do the following:

- Set up, configure, maintain, and troubleshoot hardware, software, SaaS products, cloud, on-premise, and hybrid networks, as well as other third-party tools.

- Add or remove users from a system or set up or change user roles and permissions.

- Write and publish code for websites, software, or other digital products.

- Respond to security-related communications or alerts that have an impact on the organization's operations.

- Provide end user support. Product support employees may be able to view and modify user data.

If you design for these scenarios, gather your security UX allies as soon as possible and make sure that you and your team understand the security risks involved. This can be very difficult when the technology is complex, and the use cases are new to you. You may need to build in additional time to collaborate with your security UX allies.

Keep in mind that users in high-risk scenarios may have specialized skills and many years of experience. But remember, all users, no matter how experienced, are still human.

Consider how the information in Table 4.1 might impact their security-related behaviors and talk to your security UX allies about how the user experience might be leveraged to address these risks.

TABLE 4.1 CONSIDERATIONS FOR USERS IN HIGHER-RISK SCENARIOS

Scenario	These users might...
Users are under pressure.	• Consider workarounds to save time. • Be reluctant to make security-related changes (like updating software) for fear it would cause a business disruption. • Be more likely to make mistakes or have slips or lapses.
Users are dealing with complex organizational dynamics, power structures, technology. They may also be in a highly regulated industry.	• Be required to rely on others for formal approvals or to make decisions. • Feel pressured to take shortcuts or create workarounds.
Operating in complex technology ecosystems.	• Develop hacks or less-than-secure workarounds so new technologies can talk to legacy systems. • Be responsible for older technologies that are challenging to keep updated and may have vulnerabilities that the organization simply can't address. • Be more likely to make mistakes or have slips or lapses.
Advanced technical skills plus privileged access could lead to an insider threat.	• Make a mistake that could have an organization-wide impact (or worse). • Have malicious intent.

Sustain the Collaboration

Consider developing design-led initiatives that foster communication and collaboration around the security user experience. First, there is nothing stopping the design team from forming its own cross-disciplinary security committee. There are plenty of low-commitment, informal methods for involving large groups of people. A "SecurityUX" Slack channel or other organizational watering holes can be a place where team members from all over the company can share and learn. Similarly, ongoing lunch-and-learns can highlight internal or external experts, panels, or question and answer sessions.

Shift Your Mindset

In this chapter, you learned:

- In order to improve the security user experience, your efforts need to be cross-disciplinary. Find your security UX allies.

- In fact, it's only when you combine your multidisciplinary superpowers that you can improve the security user experience.

- You need a diverse team to make the security user experience usable and accessible.

- Make sure to leverage threat modeling techniques. I recommend using Adam Shostack's four-question threat modeling framework:

  - What are we working on?

  - What can go wrong?

  - What are we going to do about it?

  - Did we do a good job?

- Ensure that everyone understands the bigger picture. The security user experience doesn't happen in insolation— it's part of a larger user journey.

- Leverage your UX visualization skills to illustrate where security might impact the security user experience.

- While it's helpful to break down the security user experience into smaller chunks, make sure to account for the fact that decisions users make in earlier parts of their user journey can impact their security later on.

CHAPTER 5

Design for Secure Outcomes

Think again about the dynamics between Alice, Charlie, and the threat actor.

Alice views Charlie as an annoying coworker who pops in at the worst times and uses language that Alice doesn't understand. So, sometimes, Alice ignores or dismisses him.

The threat actor is in the background, observing and preying on this dynamic.

Charlie could be really helpful to Alice, if only he anticipated her needs. If only he were a better, more helpful coworker. Most importantly, for this relationship to work, Alice needs to be able to trust Charlie.

Remember, in most cases, Alice expects the experience with the product to be safe. In other words, she expects Charlie to be doing his job, not bothering her.

Yet, bothering Alice is exactly what Charlie often ends up doing. Because of this, in order to truly change the dynamic—and ultimately improve security outcomes—you're going to need to get your team and your security UX allies to consider two things:

1. First and, ideally, how might we design a safer system for Alice?

2. How might we fix the dynamic between Charlie and Alice?

How Might We...?

When you think about potential security UX design solutions, think big and think bold. Most importantly, involve your cross-disciplinary team. Think about the tools from your UX toolkit you might be able to leverage.

"How might we...?" questions (a practice introduced at Procter & Gamble and used by Min Basadur) might help set the right mood for out-of-the-box thinking.[1] Remember, your UX team isn't—and shouldn't—be solving for the security user experience alone. In other words, the "we" part in "How might we...?" is really important.

---

1 Salesforce has a great article on how to craft "How might we...?" statements in "Create Effective How Might We Statements," Salesforce Trailhead (course), https://trailhead.salesforce.com/content/learn/modules/challenge-framing-and-scoping/create-effective-how-might-we-statements

The "How might we...?" questions posed in this and the next chapter are intentionally broad to cover a wide range of scenarios. Use them as a starting point so that you can develop your own questions specific to your unique circumstances.

Each "How might we...?" heading includes "Alice" intentionally. You can't anticipate how Alice will react if you aren't channeling her at every point in this process. Hopefully, this is understood among your team, but reminders go a long way. Think about opportunities to continually remind your security UX allies who you are designing for. For example, include references to your personas during workshops and presentations (Figure 5.1).

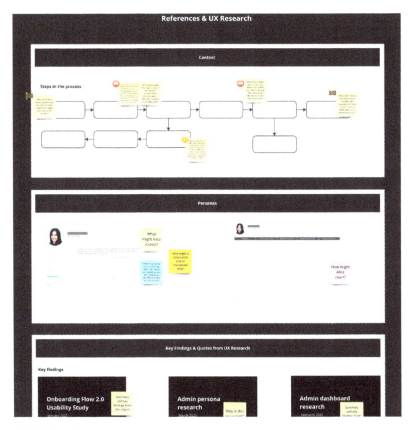

FIGURE 5.1

When brainstorming potential solutions, it may be helpful to remind your team who you are designing for and what security-relevant information to keep in mind.

How Might We Make the System Safer for Alice?

Start with a "How might we...?" question that might, at the surface, seem really obvious: "How might we make the system safer for Alice?" You design safer systems by understanding Alice, Charlie, the threat actor and how they impact one another in the security ecosystem.

But let me diverge for a moment to use a nonsecurity example to explain what I mean. Anyone who works with horses regularly knows they are accident-prone and aren't always aware of where they put their bodies. Smart equestrians keep a tidy barn area and fastidiously remove sharp objects not because they are particularly neat or well-organized. Instead, tidiness prevents horses from stepping on nails (often on the ground after a horse gets new shoes) or catching themselves on a protruding object when startled. By accounting for how horses behave, horse people have developed a system that is safer for horses—and humans.

You know Alice isn't a horse. But hopefully you see the analogy. The point is you are able to design a safer system because you understand both Alice and the risks that Alice might face.

The underlying concept here is far from new. In fact, human factors practitioners have been talking about it for a long time. Ideally, you should design the system so that bad things can't possibly happen. If you can't do that, design the system so it's extremely unlikely they will happen. As Michael S. Wogalter describes in *Handbook of Warnings*, the optimal solution will always be to "design out the hazard." If you can't do that, "guard against the hazard." The last resort is to warn the user (and hope they will comply—more about that later in this chapter).[2] Remember this as you read the rest of the chapter.

Don't be afraid to question how the features you introduce might impact the security and privacy of your users. Don't be afraid to suggest alternatives. Remember, this isn't someone else's job. You are the advocate for your users. With that in mind, you can ask things such as:

- What are the security and privacy implications of adding this feature?
- What are the security and privacy implications of gathering this data or information? (Remember, information security is about protecting information, and you don't need to bother protecting anything you don't have. Check out the sidebar, "You Don't Have to Protect Information That You Don't Have," in Chapter 6, "Design Access.")
- How might we reduce the chances of the bad thing happening?

Reducing the chances that the bad thing can happen is exactly what the coffee grinder manufacturers did in Chapter 3, "Beware of Unintended Consequence." For a digital example, Gmail knows it's difficult for Alice to distinguish a safe email from a malicious one. Because of this, if Gmail detects a malicious attachment—something it scans for automatically—Alice simply won't see it (Figure 5.2).

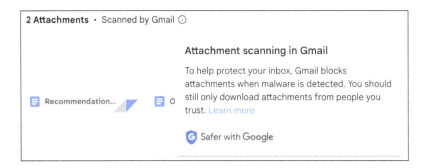

FIGURE 5.2
Gmail scans attachments for malware, and if detected, Gmail will remove the attachment, so the user never sees it.

---

2   Michael S. Wogalter, ed., *Handbook of Warnings* (Mahwah, NJ: Lawrence Erlbaum Associates, 2006), 4.

The recommendation is also echoed in a human factors textbook: Christopher D. Wickens et al., *An Introduction to Human Factors Engineering*, 2nd ed. (Upper Saddle River, NJ: Pearson Education Inc., 2004), 374.

Similarly, if Alice tries to *send* a malicious attachment, Gmail will warn her and won't allow her to attach the file.[3]

Before you start brainstorming solutions, however, make sure that you and your security UX allies take the time to uncover "what can go wrong?" (One of Adam Shostack's threat modeling questions is referenced in Chapter 4, "Find the Right People, Ask the Right Questions." If you skipped this part, go back, because you can't solve for a problem you haven't identified.) I can tell you from personal experience that cross-disciplinary collaboration will help you immensely: your security UX allies can both help you surface potential security and privacy issues, as well as being your most creative counterparts in coming up with solutions.

How Might We Leverage "Secure by Default"?

Another way you can design safer systems is by leveraging secure defaults. According to Richard Thaler and Cass Sunstein, the authors of *Nudge*,[4] many people don't change the product or system's default settings. In some cases, users won't even know security or privacy settings exist. I've spoken to many people who did not know what two-factor authentication was. Would these folks just magically find two-factor authentication in their settings? Probably not.

According to the "Secure by Design Guidelines" (guidelines developed as a result of collaboration between the United States Cybersecurity and Infrastructure Security Agency [CISA] and international agencies), secure by default refers to "products that protect against the most prevalent threats and vulnerabilities without end-users having to take

---

3   Will there be cases where Gmail is wrong? Of course. But Gmail designed the system so it is much more difficult to send or receive malware.

4   Richard Thaler and Cass Sunstein, *Nudge: The Final Edition* (New York: Penguin Books, 2021).

additional steps to secure them."[5] In addition to helping people who might not know settings exist, taking a "secure by default" approach also helps when Alice might forget to double-check security settings, or when Alice is rushing through a task and may select the wrong option.

When I started using the Zoom app (video conferencing software) on my iPad, I wanted to upload a background photo. To my privacy-minded delight, Zoom only had access to the photos I explicitly chose—nothing else. Authors James Dolph and David B. Cross use this example in "It's Time to Elevate. Moving the Discussion Forward on Secure by Default"[6] as an example of secure by default design.

In other words, iOS is protecting my photos by design, without me having to do anything. When a third-party app requests access to my photo albums, I select what I want to share, if anything (Figure 5.3). The apps don't have blanket access to the photos on my phone.

FIGURE 5.3
On iOS, the user has explicit control over what photos apps have access to. This is an example of a design being secure by default.

---

5   "Secure by Design: Shifting the Balance of Cybersecurity Risk: Principles and Approaches for Secure by Design Software," Cybersecurity and Infrastructure Security Agency (CISA) (guide), www.cisa.gov/resources-tools/resources/secure-by-design

"Fail-safe defaults" was a principle recommended by Jerome H. Saltzer and Michael D. Schroeder, "The Protection of Information in Computer Systems," *Proceedings of the IEEE* 63, no. 9 (1975): 1278–1308, https://doi.org/10.1109/PROC.1975.9939

Lorrie Cranor recommends "well-chosen defaults" in Lorrie Faith Cranor, "A Framework for Reasoning About the Human in the Loop," *UPSEC'08: Proceedings of the 1st Conference on Usability, Psychology, and Security* (April 2008): 1–15, https://doi.org/10.1184/R1/6620651.v1

6   James Dolph and David B. Cross, "It's Time to Elevate. Moving the Discussion Forward on Secure by Default," IT ISAC (whitepaper), www.it-isac.org/critical-saas-sig

How Might We Guide Alice Along the Safer Path?

Many security experts emphasize making the safer path the easier path. Jason Puglisi, an application security engineer at a well-known technology company, gives the following sage advice: "Make it hard to do the wrong thing."[8]

eero, a router manufacturer owned by Amazon, has a specific audience: everyday people like Alice who just want to watch shows, shop online, and check their email—people who don't *want* to bother setting up their router. eero knows Alice doesn't know much about setting up her home network, so they've designed eero precisely for her.

---

7   Don Norman, "Principles of Human-Centered Design," Nielsen Norman Group (video), www.nngroup.com/videos/principles-human-centered-design -don-norman

8   Voice+Code, "Security Engineers Hate CAPTCHAs, Too with Jason Puglisi," *Human-Centered Security* (podcast), November 17, 2023, https://share.transistor.fm/s/504dd7c8

Similar advice is given by Richard Thaler and Cass Sunstein, *Nudge: The Final Edition* (New York: Penguin Books, 2021).

The best part? eero made the *entire* setup experience easier—not just the security parts. Investing in this part of the user experience— setup and onboarding—has clear product *and* security ROI (return on investment). It gets Alice to the product's value faster. She gets up and running and doing the thing she signed up for or paid to do. **From a security perspective, the choices Alice makes during setup are things that could negatively impact her security later on. Guiding her along the safer path helps future-proof her security.**

While you may never design the onboarding experience for a router, it's quite likely your product involves sign-up and onboarding, where users have to go through a series of steps before they start using the product. Some of those steps likely involve making security-related decisions and taking security-related actions.

> **NOTE** SECURITY IS EMBEDDED IN THE USER
> EXPERIENCE—WHETHER ALICE THINKS ABOUT
> IT OR NOT
>
> Remember, for Alice, the security user experience is *just part of her experience*. Sometimes she's not even thinking about security or privacy. By thinking across the *entire* user experience and designing with Alice's goal in mind (getting up and running and watching her favorite streaming shows), eero has improved the entire onboarding process, including security—whether Alice is aware of the security bits or not.

How Might We Optimize Onboarding and Setup for Alice?

Imagine that Alice just moved into a new home, and she's unboxing her new eero router. Next to her is a different, but similar-looking device that her internet service provider (ISP) gave her: a modem. The last time Alice set up her home network was when she moved more than 10 years ago. She's not quite sure how to get everything set up.

eero is thoughtfully designed to help Alice through every part of the setup process. In fact, it would be extremely difficult for Alice to set up her router incorrectly.

First, eero presents timely instructions and infographics explaining the difference between a modem and a router. It even tells Alice the order in which the devices should be plugged in. It guides Alice through each minute step—steps that, if done improperly, might cause her setup to fail—and it even provides a visual of what the complete setup should look like, as shown in Figure 5.4.

FIGURE 5.4
eero anticipates that Alice may not know how to set up her modem and router and uses clear instructions and infographics to guide her through every step of the process.

While knowing what device to plug in when is not a security concern, it's a great example of anticipating what the user doesn't know and providing a helpful set of steps to improve the user experience overall.

In the past, routers had default usernames and passwords (usually something very easy-to-guess, such as username: admin and password: password) that Alice would have to know about and proactively find and change. As you can imagine, many people never changed these default passwords. They either didn't know about them, or they didn't bother.

Anticipating this, eero guides Alice through the process of enrolling her phone to receive one-time passcodes (OTPs) each time that she signs in. Alternatively, Alice could use her Amazon account credentials. Alice can't forget this step because it is part of the setup process.

Finally, eero chooses the most secure Wi-Fi settings. It anticipates that Alice probably won't know the difference between WEP, WPA, WPA2, and WPA3—and won't go through the trouble of figuring out what the differences mean or how they might apply to her. eero doesn't make Alice choose between options that Alice doesn't understand, like older routers used to do. It simply chooses the most secure one.

> **NOTE** EERO REDUCES CHANCES FOR MISTAKES, SLIPS, AND LAPSES
>
> eero's onboarding prevents mistakes, slips, and lapses. Because Alice is guided through the process, it's OK if she has no idea how to set up a router securely. eero does nearly all the work for her. In addition, by guiding Alice through the process, eero also prevents slips (where Alice might choose the less secure option because she is distracted) or lapses (where she might forget to complete an important security step, like changing default credentials).

Find What's Right for Your Users

The point of highlighting the eero onboarding process is not to prescribe a specific solution but simply to call out that you have the opportunity to improve whole parts of the user journey—such as registration and setup—with security baked in as part of it.

Perhaps you are designing enterprise software or products for a specialized, technical audience. Maybe Alice requires greater control or more advanced options. Your user journey is much more complicated than eero's. What then?

Tom Keenoy, a design leader at a cybersecurity company, designs software used by specialized power users (for example, security analysts and IT administrators). Tom's advice? "Watch your users before you try to solve a problem and look for where the holes in the process are. If you don't know what the user's process is, you can't design for it."[9] For Tom's users, that means providing the right information at the exact moment of need.

But you see a common theme: Tom is explaining that you need to understand your users—ideally by directly observing their work-flows—and anticipate where things might go wrong.

No matter who your users are—technical or nontechnical, security experts or not—no one likes to muddle through a confusing user experience. Figure out the ways in which Charlie can help Alice set up her products and accounts in the most secure way for her and her circumstances. The takeaway? Work with your cross-disciplinary team to understand what the right solution is for your users and your product. Like eero, you'll probably improve the overall user experience in the process.

> **NOTE** WHAT ARE THE UNINTENDED CONSEQUENCES?
>
> Helping users could mean making things easier—or harder. Always keep in mind that making actions easier isn't always the answer when it comes to security. And, unfortunately, when you focus on metrics like conversions, your team may over-emphasize getting the user from point A to point B faster at the expense of security.
>
> Sometimes, it's OK to slow users down. Remember the coffee grinder example—it intentionally introduced friction to keep people safe. There will be moments where you need to introduce friction thoughtfully.
>
> Keep in mind, however, that introducing friction may backfire. Dr. Margaret Cunningham (a behavioral scientist who worked at a cybersecurity company) says, "What happens when you layer a bunch of security on someone? [They find ways around it.] That's not right. We really don't want to just end up making it so we are increasing workarounds and rule-breaking."[10]
>
> Always ask your security UX allies, "What are the unintended consequences of our design decisions?"

---

9 Voice+Code, "Designing for Cybersecurity Power Users with Tom Keenoy," *Human-Centered Security* (podcast), November 29, 2023, https://share.transistor.fm/s/6dfc24db

10 Voice+Code, "We All Have Been the 'Stupid User' at Some Point with Dr. Margaret Cunningham," *Human-Centered Security* (podcast), February 10, 2021, https://share.transistor.fm/s/0d7df5af

How Might We Help Charlie Help Alice?

There are moments in Alice's user journey where she needs to make a security-related decision or take a security-related action and Charlie should be helping her (as described in Chapter 1, "Security Impacts the User Experience"). But all too often, he isn't.

These could be moments when Alice is reading and interpreting instructions, error messages, warnings, security and privacy policies, or explanations about what personal information is being collected.

The following section is intended to help you improve *any* interaction between Charlie and Alice—regardless of the context or medium.

Think back to the introduction to Chapter 2, "The Players in the Security Ecosystem," when Charlie calls Alice and interrupts her right before a holiday weekend. He uses acronyms Alice can't understand. He's throwing a bunch of extra work on Alice's plate. Alice suspects Charlie might be over-reacting just like he always does. Because of these dynamics, Alice may be thinking things like:

- "Why should I care? What does this mean? On a scale from "meh" to panic, where should I be?"
- "How do I know this is legitimate?"
- "You're telling me I need to go do something. But why? Right now? Do I have to? What happens if I don't?"
- "OK, well, what should I do?"

> **NOTE** RESOURCE: LEVERAGE HUMAN-IN-THE-LOOP FRAMEWORK QUESTIONS
>
> When evaluating an existing or potential error or warning message, first run through the questions at the end of Lorrie Cranor's human-in-the-loop framework[11] (referenced in Chapter 4). Warnings and error messages are exactly the type of scenarios the framework was designed to evaluate. For example, one question is, "What type of communication is it? Is it active or passive? Is this the best type of communication for this situation?" These questions will give you a systematic way to put yourself in Alice's mindset and understand how she might react to the warnings or errors presented to her.

---

11 Cranor, "A Framework for Reasoning About the Human in the Loop."

Alice needs to know that Charlie is there to help her—that Charlie has her best interests in mind. Alice will not be receptive to anything Charlie has to say unless you first fix her relationship with Charlie.

Help Charlie help Alice.

Alice Needs to Trust Charlie, but Recognize His Limitations

Improving the relationship between Alice and Charlie does not imply Alice doesn't need to recognize and take into account Charlie's limitations. This nuance is a unique, but important, UX challenge. As with any relationship, Alice needs to exercise her own judgment. She needs to know what Charlie's limitations are.

For example, upon signing up, ChatGPT urges users not to share sensitive information (Figure 5.5)—ChatGPT captures and learns from any information you provide it. While users may ignore or forget this warning, it's a good example of how ChatGPT has anticipated and accounted for what users might—and might not—understand about the product and its limitations and made a UX decision to help set users' expectations.

This assumes, of course, that Charlie is designed with Alice's best interests in mind. (Check out the sidebar, "When You're in a Position to Influence Behaviors, Question Your Priorities," in Chapter 2.) And it assumes that Charlie hasn't been manipulated by a threat actor—both factors your team needs to consider.

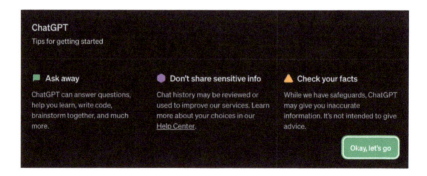

FIGURE 5.5

As part of its onboarding "tips," ChatGPT urges users to not "share sensitive info."

Choose Your Words Carefully

To improve Charlie's relationship with Alice, the words you use are critical. Carefully choose them. It's really confusing out there for users. It's hard to know what to do. The security industry is full of technical jargon and concepts that users don't fully understand (or are aware of at all).

Don't assume that Alice knows what the words mean. Words really matter in improving the security user experience. If you have content designers on your team, get them involved!

Honestly, it was challenging to find examples of organizations doing this right. But one organization stood out: HEY, an email client from 37signals. HEY provides one of the best examples of explaining terms in a way that a user can understand, while simultaneously fostering trust and reinforcing their product's value and branding.

HEY blocks spy trackers, a feature HEY anticipates a user may not know much about.

HEY answers Alice's question before she can even ask it: Why should I care? How does this impact me? Then HEY wrote a clear explanation to address Alice's questions (Figure 5.6):

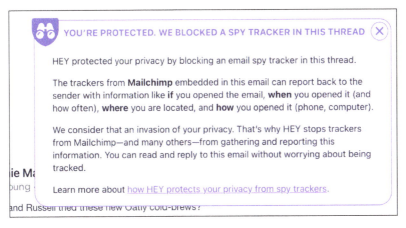

FIGURE 5.6

HEY anticipates that Alice may not know what spy trackers are. It offers a simple explanation that addresses how spy trackers impact Alice.

The trackers from **Mailchimp** embedded in this email can report back to the sender with information like **if** you opened the email, **when** you opened it (and how often), **where** you are located, and **how** you opened it (phone, computer).[12]

Also pay attention to where HEY decided to do this. It's placed in context, where Alice is most likely to see it and where it matters most: the email in which the spy blocker is being blocked.

More importantly, HEY uses the communication to establish trust and reinforce their brand. The end of the message says, "We consider that an invasion of your privacy" with a link to learn more.

NOTE THERE ARE ALWAYS TRADE-OFFS

Note that HEY's spy tracker notification is taking up a lot of real estate, getting in the way of the user's primary task: reading and responding to email. To be fair, the screenshot included here is an expanded view. The default view is much smaller. However, this is a trade-off. Your team needs to decide how important the communication is to draw attention away from the user's primary task. Speaking of which, you don't have to cram everything into one place. Consider where you can link to more information. Just keep in mind that users may not bother to read it. For this reason, ensure that the most critical information is in a place where users are most likely to see it, as researchers suggest in "Alice in Warningland: A Large-Scale Field Study of Browser Security Warning Effectiveness."[13]

Be One Step Ahead of Skepticism

Keep in mind that Alice may be skeptical of Charlie. You need to anticipate this. HEY's security page does a remarkably good job addressing one of the most common skeptical remarks many people have said to me over the years. Remarks like, "Why do I need to do [this security-related thing]? I have nothing to hide."

---

12 "Blocking Email Spies 24-7-365," HEY, www.hey.com/features/spy-pixel-blocker/

13 Devdatta Akhawe and Adrienne Porter Felt, "Alice in Warningland: A Large-Scale Field Study of Browser Security Warning Effectiveness," *Proceedings of the 22nd USENIX Security Symposium* (SEC'13) (August 2013): 257–272.

Alice may ask this very question when she is required to use 2FA.

This could be a potential roadblock for HEY onboarding because, in order to use HEY, you must have 2FA enabled after an initial trial period—something that, at this time, is not required by HEY's competitors.

HEY anticipates potential pushback and has a perfect response: "You have everything to hide [...] Email is the skeleton key to the rest of your digital life." The rest of the explanation is easy-to-understand, fosters trust, and is totally on-brand for HEY (Figure 5.7).[14]

Anticipate where Alice will be skeptical of your message. Address it head on. And build trust in the process.

Serious security
Email is the skeleton key to your digital life.

Few things are as important to protect. Here's an overview of our approach to ensure that your HEY account stays secure.

FIGURE 5.7
HEY anticipates that Alice may be skeptical of turning on two-factor authentication—something she must do in order to use HEY. HEY's security page explains why two-factor authentication for email is important and does so in a way that is easy-to-understand and serves to foster trust in the product.

---

14 "Serious Security," HEY, www.hey.com/security/

Leverage Warning and Error Message Best Practices

In their book, *Writing Is Designing: Words and the User Experience*, Michael J. Metts and Andy Welfle provide the following error message guidelines:

- "**Avoid:** Find ways to help your users without showing them an error.
- **Explain:** Tell your users what's going on and what went wrong.
- **Resolve:** Provide a solution to the problem that the user is facing."[15]

The guidelines also apply nicely not only to error messages, but also to warnings. I particularly like Michael and Andy's guidelines because they are easy to remember and apply.

These guidelines are included in the chapter called "Errors and Stress Cases" in their book, and though not security-specific, the chapter provides well-rounded advice that applies to a lot of security-related situations.

To expand on Michael and Andy's guidelines, usable security researchers at Carnegie Mellon University, who wrote "Warning Design Guidelines" (well worth reading in full), drive home a particularly important point: help Alice understand how the security risk impacts her and provide "relevant contextual information."[16]

---

15 Michael J. Metts and Andy Welfle, *Writing Is Designing: Words and the User Experience* (New York: Rosenfeld Media, 2020), 62.

   In addition, it's worth reviewing the following resources from Nielsen Norman Group:

   - Jakob Nielsen, "10 Usability Heuristics for User Interface Design," Nielsen Norman Group (article), updated January 30, 2024, www.nngroup.com/articles/ten-usability-heuristics/
   - Tim Neusesser and Evan Sunwall, "Error-Message Guidelines," Nielsen Norman Group (article), May 14, 2023, www.nngroup.com/articles/error-message-guidelines/

16 Lujo Bauer et al., "Warning Design Guidelines," Carnegie Mellon University (CMU-CyLab-13-002), February 5, 2013, https://kilthub.cmu.edu/articles/journal_contribution/Warning_Design_Guidelines_CMU-CyLab-13-002_/6468131

Here's how these guidelines apply to a recent notification I got on my phone. It said "Your current location can be seen by the owner of this AirTag." AirTags are small metal discs you can use to track your keys or other objects. But they can be used to track people, too.

The notification helped me in a couple ways:

1. It helped me understand what it meant to have an AirTag nearby—particularly useful if I didn't know what an AirTag was.

2. It provided clues that helped me determine if this was something I needed to be worried about or a false positive (something that might be malicious, but in the context of the situation, is not). In other words, it provided evidence. It was specific. It even provided a photograph of what an AirTag looked like and indicated when it started tracking me (Figure 5.8).

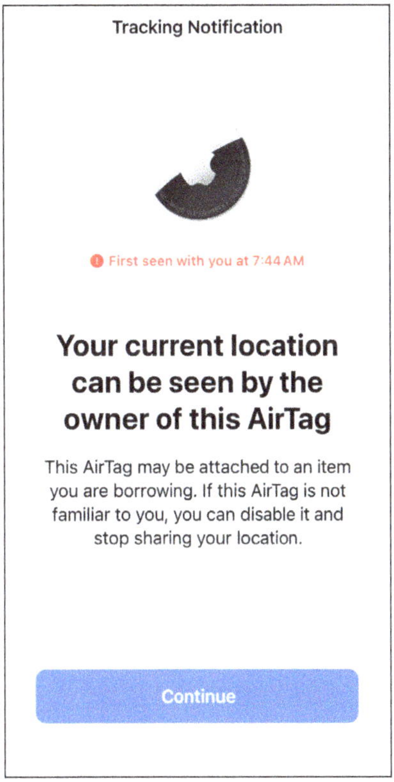

There was a perfectly reasonable explanation for why this AirTag was tracking me—I borrowed a set of keys with an AirTag attached to them. All of these clues can help me determine how to proceed—all delivered without unnecessary drama. The lack of drama is important: it helps foster trust between Alice and Charlie.

Tracking Notification

🔴 First seen with you at 7:44 AM

Your current location can be seen by the owner of this AirTag

This AirTag may be attached to an item you are borrowing. If this AirTag is not familiar to you, you can disable it and stop sharing your location.

Continue

FIGURE 5.8
Apple devices warn users when they detect an AirTag may be tracking their location, which helps you figure out if there is a legitimate reason an AirTag is near you.

Lean on the Technology

Nothing is more frustrating than knowing something is or might be
wrong and having no idea how to fix it.

In other words, it's not enough for Charlie to say, "Hey, Alice, there's
a problem." Charlie needs to help Alice fix it. This recommendation
isn't new: it's in "Warning Design Guidelines" referenced in the
previous section: "offer meaningful options [...] and enough information to decide between them."[17]

But sometimes it's hard to know what to tell Alice because often the
answer is "it depends." That is not an excuse to let Alice figure things
out for herself.

Instead, ask your team, how might we leverage technology to help
Alice in this situation?

Imagine that Alice is a developer using GitLab. One of the things
GitLab does is help Alice find and prioritize vulnerabilities in her
code. GitLab introduced a feature called *GitLab Duo Chat* that uses
AI to help Alice understand the vulnerability, how the vulnerability

---

17 Bauer et al., "Warning Design Guidelines."

might impact the product and its particular circumstances, and even provides options to help Alice fix the problem.[18] See Figure 5.9.

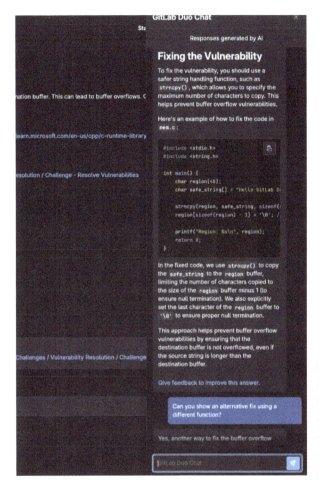

FIGURE 5.9
GitLab Duo Chat leverages AI to help users better understand a vulnerability in the code and how to fix it, as shown in a recent GitLab blog post.

The best part: GitLab didn't just anticipate that Alice will have questions—it allows her to ask them in real time using her own words. GitLab is leaning on technology to help Alice in the context of her own situation.

18 Alana Bellucci and Michael Friedrich, "Developing GitLab Duo: Use AI to Remediate Security Vulnerabilities," *GitLab* (blog), July 15, 2024, https://about.gitlab.com/blog/2024/07/15/developing-gitlab-duo-use-ai -to-remediate-security-vulnerabilities/

Note that with AI (and the introduction of any new technology) comes unintended consequences, like over-reliance on the technology. Check out the sidebar "When You're in a Position to Influence Behaviors, Question Your Priorities" in Chapter 2 for more information and resources.

Don't Contribute to the Confusion

When it comes to the security user experience, lack of consistency—and unnecessary added confusion—is your worst enemy. In fact, "consistency and standards" is one of Jakob Nielsen's 10 Usability Heuristics.[19] This heuristic very much applies to the security user experience.

Different URLs used for different circumstances, inconsistent messaging, inconsistent visuals—all of these things make it harder for users to distinguish real (a legitimate communication coming from you) from fake (a threat actor posing as your company or product).

I worked with an organization that required users to go through an annual compliance process, which meant that users needed to visit a website and answer some questions. So, once a year, users would get an email about a process they had long forgotten about, from an email address that was inconsistent with other communications from the company. The email asked them to go to a URL they'd never been to before with a visual design inconsistent with the rest of the product or marketing site. The website asked for sensitive information, and, not surprisingly, many users simply failed to complete the process, thinking they might be on a malicious website. That's good for them! They were right to be suspicious.

How do I know this? It was in our support transcripts. Take that as a warning: inconsistent or unclear communications can cause a flurry of support calls and angry and confused customers. Don't do this to Alice. You are burdening her unnecessarily. And you are setting her up to fail.

---

19 Nielsen, "10 Usability Heuristics for User Interface Design."

PSYCHOLOGY AND BEHAVIORAL ECONOMICS
RESOURCES

There are a wealth of psychology and behavioral economics
resources that will come in handy when you start thinking about
designing for secure outcomes. Here are a few that have been
helpful to me:

- *Nudge: The Final Edition* by Richard Thaler and Cass
  Sunstein.[20]
- Speaking of which, "Nudges for Privacy and Security: Under-
  standing and Assisting Users' Choices Online"[21] is a primer on
  nudges applied directly to the security user experience.
- *Start at the End: How to Build Products That Create Change*
  by Matt Wallaert.[22]
- Nielsen Norman Group's "Psychology for UX: Study Guide."[23]

20 Thaler and Cass Sunstein, *Nudge: The Final Edition*.

21 Alessandro Acquisti et al., "Nudges for Privacy and Security: Understanding
   and Assisting Users' Choices Online," *ACM Computing Surveys* 50, no. 3 (2017):
   1–41, https://doi.org/10.1145/3054926

22 Matt Wallaert, *Start at the End: How to Build Products That Create Change*
   (New York: Portfolio/Penguin, 2019).

23 Tanner Kohler, "Psychology for UX: Study Guide," Nielsen Norman Group
   (article), January 10, 2024, www.nngroup.com/articles/psychology-study-guide/

Shift Your Mindset

In this chapter, you learned:

- First, consider how you can design the system so that the bad thing can't happen.

- Leverage the principle of "secure by default" to help keep Alice safe.

- Don't just let Alice figure security out for herself. Guide her along the safer path—this helps prevent mistakes, slips, and lapses.

- Figure out how Charlie can best help Alice. Be mindful of the words you use—it's confusing out there.

- Leverage warning and error message best practices. Michael J. Metts and Andy Welfle say it best: "Avoid, explain, resolve."

- In fact, telling Alice there is a problem simply isn't enough— Charlie needs to help Alice fix it. Technology can help.

- Keep in mind that there is not a one-size-fits-all solution. You need to find what's right for your Alice.

Design Access

The underlying concept of access control is ingrained in us. We instinctively practice access control with anything we value. We lock our cars, stash family heirlooms in a safe, and put that special chocolate bar in the upper cabinet where the kids can't reach it.

Even squirrels stash away their nuts wherever it is that squirrels keep the things most precious to them.

We limit who has access to our possessions—our car, our valuables, or our candy (or nuts)—by making access more difficult and effortful, and less convenient. We do this because we know you can't steal or damage what you don't have access to.

The same goes for information.

"Access control" or "access management,"[1] as security folks refer to it, is a big topic and whole books are devoted to it. (I recommend checking out Paul van Oorschot's *Computer Security and the Internet: Tools and Jewels from Malware to Bitcoin*.[2]) I've only scratched a tiny bit off the surface of a huge—and hugely important—domain. Two things I can tell you, though:

- **Access control is really, really important.** You can't protect information or information systems without controlling access to them.

- **It's challenging to balance access control with usability.** Jared Spool coined the term *selective usability*, and it perfectly describes the problem with access control.[3] You want to allow legitimate users access to their information anywhere at any time, yet you don't want to give unauthorized users the same access.

I'm going to use the example of Figma for a portion of this chapter to illustrate some basic concepts around access.[4]

Along the way, I'm going to highlight why certain choices may have been made. (Note: I'm not affiliated with Figma, so these are my own assumptions.) As you read through these sections, know that access is "designed" just like any other part of the user experience.

---

1   In the digital realm, people are not the only grantees of access. Hardware, software, and digital processes and services can also be granted access. For the purposes of this book, however, the primary focus is on people—your users—getting access to information.

2   Paul van Oorschot, *Computer Security and the Internet: Tools and Jewels from Malware to Bitcoin*, 2nd ed. (New York: Springer, 2021).

3   Jared Spool. "Insecure and Unintuitive: How We Need to Fix the UX of Security," UX Immersion: Interactions (presentation), May 2017, www.uie.com/uxsecurity

4   Figma can get pretty complicated, and it depends on what plan you have, so the examples have been simplified for clarity.

YOU DON'T HAVE TO PROTECT INFORMATION THAT YOU DON'T HAVE

The less information you gather, the less information you have to worry about protecting. A disgruntled employee can't sell information your organization doesn't have. Hackers can't access information you don't have. This reduces the risk to everyone—your users and your organization.

Don't be afraid to ask your team:

- Do we have to capture or ask for all of this information?
- Is it really necessary?

In fact, these types of questions are precisely what Debbie Reynolds, privacy expert and host of *"The Data Diva" Talks Privacy* podcast, urges teams to ask. Debbie talks extensively about the challenges that organizations face trying to comply with laws, protect personal information, and still provide a good user experience. Her content is well-worth following.[5]

Further, with the proliferation of the use of artificial intelligence (AI), big tech companies are asking similar questions. IBM's ethical AI principles and questions to consider include: "How do we create the best user experience with the minimum amount of required user data?"[6]

Google's "Responsible AI Practices" advises, "If it is essential to process sensitive training data, strive to minimize the use of such data."[7] Know that you are in good company when asking similar types of questions of your team.

One more resource worth checking out: in the blog post, "Formalizing an Ethical Data Culture," Gwen Rino, senior data scientist from Code for America, outlines a set of guidelines for team members to leverage for gathering, protecting, using, and interpreting data. The post links to the "Ethical Data Use at Code for America" guidelines. What's particularly powerful is the emphasis on questioning why you are capturing the data in the first place. Can you develop similar guidance for your team?[8]

---

5   A great starting point is the Debbie Reynolds Consulting LLC YouTube page, with many of Debbie's informative videos, as well as *"The Data Diva" Talks Privacy* podcast at www.youtube.com/@debbiereynoldsconsultingll8529/featured

6   IBM Design for AI, "AI Design Ethics Overview," IBM, www.ibm.com/design/ai/ethics/

7   Google AI, "Responsible AI Practices," Google, https://ai.google/responsibility/responsible-ai-practices/

8   Gwen Rino, "Formalizing an Ethical Data Culture," *Code for America* (blog), April 15, 2020, https://codeforamerica.org/news/formalizing-an-ethical-data-culture

How Might We Design Access for Alice?

Imagine that Alice is the head of UX at a digital design firm. Alice and her team use Figma to collaborate on, share, and store design files. Figma operates in a similar way to how you can create, edit, collaborate on, and share files in Google Workspace or Microsoft 365.

As most design teams can attest, the beauty of cloud-based design tools is the ease of collaboration they provide. Design files are in one place and can be shared with and edited by multiple people in real time.

Everyone on Alice's team has their own account, but only Alice can add and delete people from the organization, view the organization's invoices, or delete the account altogether. Alice is the only one authorized to take these actions.

Alice, the head of UX, is authorized to add a new designer, Bob. But she doesn't want Bob to be able to add new users. In other words, Alice needs to be the administrator, while Bob needs to have a role with fewer privileges.

Figma, similar to many other cloud-based services, accomplishes this through role-based access control (RBAC)—something that is very common in SaaS software. Users are assigned to a group and each group has a set of privileges. WordPress does something similar, but their setup is more straightforward, which is why I'm using it as an example here. You can see WordPress roles (as shown on WordPress's support page) are "administrator, editor, author, and contributor" (Figure 6.1).[9]

FIGURE 6.1

Here are role-based access controls in action. WordPress has roles for an administrator, editor, author, and contributor.

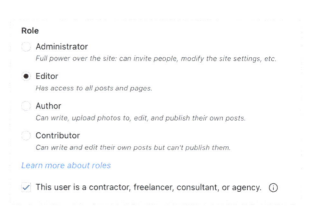

---

9   "User Roles," WordPress Support, https://wordpress.com/support/invite-people/user-roles/

For Access, Authorization Is Like a Design System

Authorization refers to what users are allowed to do in certain contexts. In other words, what privileges or permissions[10] they have when they use the system. This can be relatively straightforward. For example, anyone can visit Figma's marketing site to view its contents, but they can't modify anything. Know that authorization put into practice can also be pretty complex: certain users can view or modify certain information in certain contexts (such as during company hours on a company-issued device).

Authorization is the security-equivalent of a design system. It's the foundation of access, and, like design systems, it can get pretty complicated. One key similarity: no one other than the design team really appreciates the work that goes into it.

There's an underlying, but important principle related to authorization: the principle of least privilege. So, let's focus on that for a moment.

The Principle of Least Privilege

The *principle of least privilege* means that users should only be able to see and do what is absolutely necessary for them to do their job or accomplish a goal. This concept is so important, it is probably engraved on a plaque in a security practitioner's office somewhere.[11]

The barista using the point-of-sale system doesn't need access to the coffee shop's financial information—they should only have access to what they need to place orders and take payments. Your doctor needs to know what medications you're currently on, but they don't need the same information about every patient at the hospital: they only need the information necessary to treat their own patients.

Here's why the principle of least privilege is important. Alice, as you learned in Chapter 2, "The Players in the Security Ecosystem," is—like all of us—human. She might accidentally share information with

---

10 I've heard people use "permissions" and "privileges" interchangeably—even security experts. There is, technically, a difference. But most people would have a hard time telling you what that difference is.

11 "The "principle of least privilege" was a principle recommended by Jerome H. Saltzer and Michael D. Schroeder, "The Protection of Information in Computer Systems," *Proceedings of the IEEE* 63, no. 9 (1975), 1278–1308, https://doi.org/10.1109/PROC.1975.9939

people she isn't supposed to or didn't intend to. In fact, a threat actor might even trick Alice into disclosing information. Or she might modify customer information or an important setting accidentally.

Or Alice may not have the best intentions. She might steal customer information and sell it to others. Or she might change an important setting with the intent of sabotaging the system.

The intention behind the principle of least privilege is to limit what the user is authorized to do. Because of this, it also limits the damage that users can cause—whether intentionally or unintentionally. Remember, if you can't access it, you can't harm it.

Now back to authorization and some new terms: *identification* and *authentication*.

Identification and Authentication

For authorization to work, you have to be able to distinguish who's who within the system, which is where identification and authentication come into play.[12]

Identification refers to the way the system distinguishes Alice from Bob from every other user in its database. Figma uses email addresses to tell users apart, so for Alice, it's "alice@email.com" and for Bob, it's "bob@email.com." Your product might use an email address, username, phone number, device ID, account number, or order number, for example.

When Alice created the Figma account, she not only told Figma her email address, but she also chose a password to associate with her email address. That password is known as *a method of authentication* or *authentication factor* (the "factor" in multifactor authentication). *Authentication* refers to evidence you provide that you are, indeed, "Alice," "Bob," or "unicornsandrainbows." In the security world, a password is referred to as *something you know*.

Authentication methods aren't limited to passwords. Alice might demonstrate that she has something like a passkey or a hardware security key (for example, a YubiKey), which is referred to as

---

12 Hat tip to Dave Johnson, Paul van Oorschot, and Jarret Raim for (ever so patiently) helping me understand the finer points around access control. Any mistakes are mine.

something you have. Or Alice might authenticate using biometrics like her fingerprint or face. This is referred to as *something you are.*

NOTE AUTHENTICATION METHODS WILL EVOLVE

As of the writing of this book, passkeys—an alternative to pass-words—are gaining traction. Authentication methods can and will evolve in the future.

Figma gives Alice the option to leverage more than one authentication method to keep her account secure. Your financial institution, for example, may require two-factor authentication. Organizations like your financial institution make two-factor authentication mandatory because there are greater risks and consequences in someone getting access to your savings account compared to, say, your digital grocery list (or they are required to do so by rules, regulations, or contracts).

In Figma's case, Alice can sign in using her password (the first authentication method), and then she'll be asked for a unique code generated by an authenticator app (the second authentication method). This is known as *two-factor authentication* (2FA) or *multifactor authentication* (MFA). As described in Chapter 3, "Beware of Unintended Consequences," 2FA is when two factors are required to sign in. MFA means two or more factors are required to access a service. They are often used interchangeably and, as far as Alice is concerned, they mean the same thing.

Design for Access-Related Scenarios

Imagine Alice's team works with freelance designers and external development firms. They also frequently have clients review works-in-progress and Figma prototypes.

When working with freelance designers, Alice wants them to be able to view and edit projects and files, but Alice doesn't necessarily want to add them to her Figma organization. That would also mean she'd have to pay for their account. Instead, Alice can simply share the file with the freelancer, and the freelancer, bob@email.com, can use his own Figma account (Figure 6.2) to access the one design file, but can't access any other projects or files in Alice's organization's Figma account.

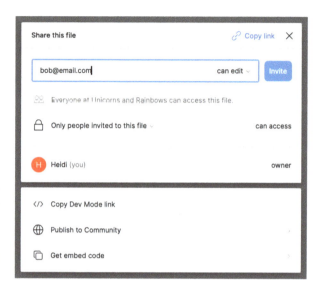

FIGURE 6.2
Alice can share select
files with someone
external to their
organization, such as
a freelancer. The free-
lancer is authorized to
access the file Alice
shared with her, but
nothing else at Alice's
organization.

Now what if Alice wants to share Unicorns and Rainbows Design
with a client? Her client probably doesn't have a Figma account and
probably doesn't want to create one. In this scenario, Alice can share
a link to the Figma prototype and set it to view only (Figure 6.3). In
other words, clients can look, but they can't touch.

Because the risk is relatively low in this scenario, identification and
authentication are kind of rolled into one. Whoever possesses the
link can access the Unicorns and Rainbows Design prototype.

The takeaway? Authorization allows you to fine-tune what users are
allowed to do and in what context.

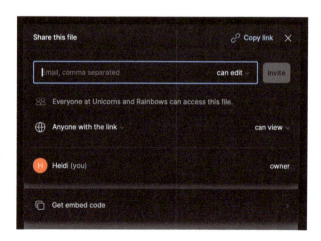

FIGURE 6.3
When Alice shares a
Figma prototype, she
can allow people who
have the link to view
the prototype without
needing to sign in. In
this scenario, clients
can look at the proto-
type but not edit it.

For FigJam, Figma's equivalent of a digital whiteboard, Alice and her team can collaborate with people who don't have a Figma account, and their access expires automatically after 24 hours. These folks can interact with and edit the board the same way that someone with an account can. They just have a limited period of time to interact. Better yet, that time is automatically set—24 hours—so Alice doesn't have to worry about remembering to disable the link at a later time (Figure 6.4).

FIGURE 6.4
If you want to be able to allow access to a FigJam file for a limited time—say, during a workshop—you can give people the capability to edit a file for 24 hours. After that, they automatically lose the ability to edit.

Account for Higher-Risk Scenarios

Talk to your security UX allies about scenarios where a user is operating in a mode where they introduce greater risks. For example, they might be able to access sensitive customer information or be able to manage or delete customer accounts. In some cases, their actions can take down entire systems and completely disrupt business operations. In other words, the consequences of their actions could impact the organization and its users.

Know that privileges can be "locked away" until you actually need them. As with the FigJam example, you can put a time limit around when someone can access a file, certain information, or perform certain actions.

Ideally, this is done automatically, without the user needing to remember anything. In FigJam, the link expires after 24 hours—there is no need for Alice to do anything.

Here's a real-world analogy: operating in a mode where you have greater privileges than you need is like carrying around your Social Security card and birth certificate with you at all times. You increase the likelihood of losing the documents or having them stolen. And having them with you constantly is completely unnecessary. Instead, you might need the documents in very rare circumstances, such as when you're applying for a passport. Otherwise, they should be locked away in a safe place.

For this reason, ask your security UX allies these questions:

- What does the user absolutely need to be able to do in order to accomplish their goal?
- Do they need to have this privilege all the time? Or just in certain circumstances?
- What other controls can we put in place when they are operating in scenarios that introduce heightened security risks?

Understand the Access-Related Tension

Imagine if Alice was not allowed to share any Figma files with anyone outside of her organization. To get around this, Alice decides to create a free account with her personal email address. She then uses the free account to collaborate with others. Worse, over time, the waters get a little muddy, and Alice and other employees are using both their work and free accounts to collaborate on projects with one another.

What Alice and her colleagues are doing is what James Reason (referenced in Chapter 2) refers to as *intentional violations*.[13] From Alice's perspective, her actions are both relatively harmless and necessary to get her job done. In fact, many people—maybe even you—have done something similar. Alice is trying to collaborate, and, from Alice's perspective, the tool isn't allowing her to do that. But her actions don't make her security team very happy. Alice shouldn't be using free or personal accounts that the company can't manage.

Dr. Calvin Nobles, portfolio vice president and dean, School of Cybersecurity and Information Technology at University of Maryland, explains, "When something becomes too complicated and it exceeds the cognitive capability of the user, users will look for shortcuts or workarounds, or they will find the most simple way to achieve their objective. That's a natural reaction, a natural instinct."[14]

In their talk "An Economic View of Usable Security," researchers Angela Sasse and Adam Beautement say, "If no one is using it properly, you're going to have bad security."[15]

You probably have hundreds of accounts you need to manage between your work and personal life. Have you ever thought to yourself, why is this all so difficult? You're not alone. There's a tension: a tension between security and the user experience. And many of the security professionals I've spoken to have felt this most acutely around access—even if they use different words to describe it.

My advice? Take the time to understand your security team's priorities and the immense amount of pressure they feel in their roles.

Remember: access control is vital to protecting information and information systems. Yet, it's challenging to balance access control with usability (remember Jared Spool's term *selective usability*).

There is no easy solution to this problem, but there are a couple things for you to think about.

---

13 James Reason, *Human Error* (Cambridge: Cambridge University Press, 1990), 195–197.

14 Voice+Code, "What Can We Learn from Human Factors Programs in Other Industries? With Dr. Calvin Nobles," *Human-Centered Security* (podcast), January 27, 2021, https://share.transistor.fm/s/1eb7c83c

15 Angela Sasse and Adam Beautement, "An Economic View of Usable Security," Microsoft (presentation), July 12, 2010, https://microsoft.com/en-us/research/video/an-economic-view-usable-security/

Learn from Your Users and Their Scenarios

One important takeaway from the Figma example in the previous sections is that Figma's team obviously accounted for the different scenarios its users (like you) might be in. For one, designers and design organizations need to be able to collaborate with team members as well as people outside their organization—people who might not have a Figma account.

In other words, Figma learned from its users, and it designed access around those scenarios: leveraging the security concepts of identification, authentication, and authorization to support those scenarios.

Here's another example of product teams that accounted for how real people use their product. Alice is at the Grand Canyon and wants to take a family photo. The only problem? A selfie just won't do—this needs to be a proper family photo. Alice spots an unsuspecting tourist, but she'd prefer not to give this stranger access to her unlocked iPhone.

This is an authorization problem—no one other than Alice should be authorized to use her phone. Except today, when Alice really, really wants to have a nice family photo taken. Apple accounted for this scenario. Anyone can take a photo or video using an iPhone while it is locked by pressing the camera icon on the bottom right of the locked home screen.

In other words, Alice can give a stranger the ability to take photos using her phone while the phone is locked, but they won't have full access to the rest of the apps and information on her device.

This scenario may seem, on the surface, pretty simplistic. But pay attention, and you'll encounter these types of access issues (and sometimes novel solutions) with just about every technology you use. By giving limited access by accounting for realistic scenarios that your users find themselves in, you can thoughtfully design access in ways that are more human and user-friendly.

There Isn't Just One Way to Design Access

There isn't just one way to design access and now, more than ever, there are options. Options that are more inclusive, accessible, and privacy-preserving.

As of the writing of this book, there are shifts happening that would give users more control over their digital identities and what information they share with different organizations. This evolution, while slow-moving, has the potential to improve both security and privacy.[16]

Bethany Sonefeld, design manager at access solutions company Duo (owned by Cisco) says, "In designing for security, we have to be mindful of what the right amount of friction is. It can't be frictionless—that's the reality. However, I do think designers [often put] the onus on the users to do the right thing."[17]

Bethany advises UX teams to work with their security UX allies to find opportunities to make the experience "a little less painful. For example, by allowing users to authenticate using passkeys or biometrics instead of a password."

For example, it's challenging to enter a passcode on a tiny Apple Watch face. Apple anticipated this, and it allows you to unlock your watch when you unlock your nearby iPhone (Figure 6.5).

FIGURE 6.5
Instead of entering the passcode on your Apple Watch, you can unlock it at the same time that you unlock your nearby iPhone.

---

16 Check out Kim Hamilton-Duffy, Ryan Grant, and Adrian Gropper, "Use Cases and Requirements for Decentralized Identifiers," W3C Working Group Note, March 2021, www.w3.org/TR/did-use-cases/

Also check out Kaliya Young, author of *Domains of Identity* (New York: Anthem Press, 2020) and on Voice+Code, "Using Self-Sovereign Identity as the Foundation for Secure, Trusted Digital Relationships with Kaliya Young," *Human-Centered Security* (podcast), December 23, 2020, https://share.transistor.fm/s/1518fce2

17 Voice+Code, "Designing Multi-Factor Authentication with Blair Shen and Bethany Sonefeld," *Human-Centered Security* (podcast), October 19, 2022, https://share.transistor.fm/s/52f79a3b

The FIDO (Fast Identity Online) Alliance is an organization that advocates for building in flexibility when it comes to authentication. It may be very difficult, if not impossible, for some users to take the necessary actions your product requires to secure their accounts and devices.

QR codes, for example, are often used when a user is setting up two-factor authentication via an authenticator app. But the FIDO Alliance's "Guidance for Making FIDO Deployments Accessible to Users with Disabilities"[18] (a must-read if you are working on authentication), advises, "QR code scanning can be challenging for users with visual disabilities. Such users may find it difficult to locate a QR code and position a camera to scan the code. QR codes can also be difficult for people with physical disabilities because it requires the user not only to hold the camera, but also to hold it steady."

Accessibility and inclusivity for two-factor authentication and passwords are also addressed in recently updated WCAG guidelines (2.2 as of the writing of this book). As these guidelines are updated over time, I suggest reviewing the latest WCAG guidelines as you collaborate with your security UX allies.[19]

Speaking of which, you will need to collaborate closely with your security UX allies (sound familiar by now?). With every option comes unintended consequences. Remember, with anything new you introduce, the threat actor will find ways around it or use it to their advantage. And Alice may do things you didn't anticipate. In fact, for this very reason your security UX allies may be reluctant to explore access-related solutions. After all, with options comes risk.

Your job is to help your security UX allies understand your users' scenarios. Help them envision themselves in Alice's situation. Help them understand what Alice might be doing right now and how she might be doing it, what technology she might be using, how she might interpret a message, or what action she might take next. And leverage resources like Microsoft's Inclusive Design Toolkit[20]

---

18 "White Paper: Guidance for Making FIDO Deployments Accessible to Users with Disabilities," FIDO Alliance (white paper), October 13, 2022, https://fidoalliance.org/white-paper-guidance-for-making-fido-deployments-accessible-to-users-with-disabilities/

19 "Success Criterion 2.2.5 Re-Authenticating," Web Content Accessibility Guidelines (WCAG) 2.2, October 5, 2023, www.w3.org/TR/WCAG22/#re-authenticating

20 "Microsoft Inclusive Design," Microsoft, https://inclusive.microsoft.design/

and the FIDO Alliance's "Guidance for Making FIDO Deployments Accessible for Users with Disabilities." Chapter 7, "Learn and Iterate," provides some additional information and resources.

Remember, there isn't just one way to design for access. By helping your UX security allies understand Alice, you can collectively design access better for her.

WHO ARE YOU EXCLUDING BY REQUIRING USERS TO CREATE AN ACCOUNT?

Consider who you are potentially excluding when you make a decision to require users to create an account. More and more services are available online, including being able to enroll in or access certain public benefits programs like Medicaid or SNAP. Code for America's "The Benefits Enrollment Field Guide"[21] refers to requiring users to create an account as a "locked front door," which "forces clients to complete a series of complex actions. Many clients cannot even look at the application without first:

- Owning and confirming an email address
- Submitting CAPTCHAs
- Setting security questions
- Creating and remembering a complex password
- Clicking on a link sent to their email account"

Instead, Code for America recommends, "At a minimum, states should offer a 'guest' enrollment that doesn't require registration."

Combatting fraud and spam are some of the reasons that users are required to create accounts. (The other main reason is marketing.) Obviously, fraud and spam are legitimate concerns, and I'm not recommending a particular solution.

However, I do recommend you talk to your security UX allies about whom you might be potentially excluding by requiring users to create an account. How might you make the experience more accessible and inclusive?

---

21 "The Benefits Enrollment Field Guide," Code for America, https://codeforamerica.org/explore/benefits-enrollment-field-guide/

Shift Your Mindset

In this chapter, you learned:

- *Access* is really important for security, but it's really difficult to balance access and usability. Jared Spool aptly coined the term *selective usability*, which describes the situation perfectly.

- *Authorization* refers to what users are allowed to do in certain contexts.

- The *principle of least privilege* means that users should only be able to see and do what is absolutely necessary for them to do their job or accomplish a goal.

- *Identification* is how the system distinguishes between Alice and Bob. It could be an email, username, account number, or any other unique representation of users.

- *Authentication* is evidence Alice provides that she is, in fact, the same "Alice" in the system.

- Learn from your users and their scenarios. Design access around this. Build in flexibility and be sure to involve people with disabilities as part of your research (described in more detail in Chapter 7).

CHAPTER 7

Learn and Iterate

The UX Working Group at the FIDO Alliance faced a challenge. Passkeys are new, so they needed to understand how to introduce people to them. Kevin Goldman explains, "We tried different moments in the journey to prompt people to create a passkey, different messaging, we iterated, and that's when we said, 'We have to stop the project.'"[1]

Kevin is the chair of the User Experience (UX) Working Group at the FIDO (Fast Identity Online) Alliance. He, his fellow UX Working Group members, as well as a cross-disciplinary team of FIDO Alliance staff and members (hereafter collectively referred to as the *FIDO Alliance team*), are working on defining and improving the user experience of passkeys.

First, a quick detour to talk about passwords and passkeys. Everyone hates passwords. But yet, so many of the digital services that people need to access rely on them. The intention behind passkeys is to make authentication easier and more accessible—*as well as* more secure. The FIDO Alliance provides technical specifications and guidance for implementing hardware security keys and passkeys—authentication methods that can replace passwords and, additionally, foster experiences that are both more usable and resistant to phishing.

Here's how passkeys work: you want to check your bank balance. You go to your bank website and then use the same authentication method you already use to unlock your device. If you're on your phone, maybe you flash your face in front of the screen or enter your phone's password. If you're on your computer, you might use your fingerprint. That's it. You're signed in.[2]

Now back to FIDO Alliance team. People of the world can't agree on much, but nearly everyone hates passwords. Because of this near-universal negative sentiment, wouldn't you expect people might jump at an alternative, given the opportunity? That's exactly what the FIDO Alliance team expected: users would sign into an existing

---

1 Voice+Code, "Learning and Iterating Are Key to Improving the Security User Experience with Kevin Goldman," *Human-Centered Security* (podcast), February 7, 2024, https://share.transistor.fm/s/c00cf8ce

2 This is an extremely simplified explanation of passkeys. To learn more about passkeys, check out: Eben Carle, "Ask a Techspert: What Are Passkeys?," *Google* (blog), October 10, 2023, https://blog.google/inside-google/googlers/ask-a-techspert/how-passkeys-work/

account with their username and password and see a message about an alternative to passwords: passkeys. The expectation was that users would see the benefit and immediately go through the simple process of setting up passkeys for their account.

Unfortunately, it wasn't quite that simple. Although people don't like passwords, signing in without them was a foreign idea.

The team knew they were missing something. But what?

Embrace Iteration, or Else None of This Will Work

After taking a step back and conducting usability studies, the FIDO Alliance team uncovered something important. Some people are familiar with passkeys because they've seen them introduced through services they use—such as Google or TikTok accounts. But, if someone is *not* familiar with passkeys and encounters a prompt to set up passkeys right after they sign in, they view it as an unwelcome interruption to their main task.

In these situations, the user simply wants to get something done: make an appointment, buy groceries, or get reimbursed by their insurance provider. They don't want to be bothered to create a passkey at this particular moment. Kevin explains, "People saw the prompt to create a passkey and thought, 'I don't know what this is,' and the majority of people new to passkeys didn't want to create passkeys right after they signed in" as illustrated in Figure 7.1.

So, what are the ideal moments to ask users to create passkeys? By taking a step to really understand the problem, the FIDO Alliance team landed on a critical insight: instead of interrupting users as they were trying to accomplish a task, the optimal opportunity to ask users to set up passkeys was during what they refer to as *account-related moments*.

These are moments when users are signing up for a new account or when they are making edits to an existing account (say, for example, when they want to change their password). Over time, as more people are familiar with the use of passkeys on various services, there will be many more places where service providers can prompt users to create a passkey.

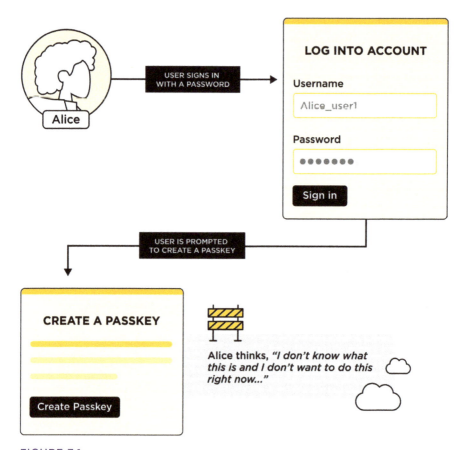

FIGURE 7.1

The FIDO Alliance team originally thought an ideal time to ask a user to create a passkey for the first time was right after they logged in using a password. But they weren't getting the adoption they expected.

By taking the time to run usability research and observe people using passkeys, the FIDO Alliance team uncovered another critical insight: users are more willing to set up passkeys during the account recovery process. In other words, when users forget their password and go through the steps to reset it, they are more likely to be receptive to the option of creating a passkey—an alternative to the password that they just forgot.

Equipped with this knowledge, the FIDO Alliance team could focus their attention on helping users adopt passkeys when users would be most open to doing so, as shown in Figure 7.2.

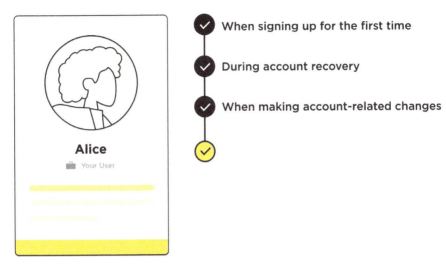

FIGURE 7.2

If people are unfamiliar with passkeys and you ask them to create a passkey right after they sign in, you are disrupting their task. Better alternatives include when Alice is signing up for the first time, during account recovery (such as when Alice forgets her password), or when she's making account-related changes. In these scenarios, Alice is more receptive to setting up passkeys.

The takeaway? Embrace iteration.

The FIDO Alliance team would not have uncovered these insights—UX roadblocks that might have prevented people from adopting passkeys—had they not paused the project to run usability research to really understand the problem. And guess what? As of this writing, the FIDO Alliance team is planning for their next round of research. In other words, the human-centered security process is never done.

In fact, as Kevin emphasizes, "Over time, the general public will be more aware of passkeys and [prompting them] to create one after signing in with a password will be more effective." In fact, there are some changes in the works that will make it easier for users who are currently using password managers (like 1Password, for example) to transition more easily to passkeys.

Kevin's point connects back to Chapters 2, "The Players in the Security Ecosystem," and 3, "Beware of Unintended Consequences,"—every change in the ecosystem affects everything else. Everything is dynamic. What we learned yesterday might not be effective six months from now.

If you don't embrace iteration, you cannot possibly improve the security user experience. And you certainly can't improve security outcomes. But, as you know, you need to rely on UX research to inform meaningful UX changes. A primer on UX research is beyond the scope of this book, but I've included some helpful resources at the end of the chapter. The following sections, however, focus on potential challenges you might face when conducting UX research related to security, as well as some recommendations on how to address those challenges.

Security User Experience Research Challenges

I had three "ah-ha" moments when considering how to conduct research to better understand the security user experience.

First, I tried to separate out security (or privacy, or safety) from the rest of the user experience. It just doesn't work that way. From Alice's perspective, she doesn't have a "security user experience"—she just has an experience. And that experience might be great or not-so-great, and security may or may not have played a role. When you have a broader lens—like observing a user on the pathway to achieve a goal—you'll get better insights. When you try to isolate security from the rest of the user experience, you get hollow, generic results.

Second, users are not going to use the same vocabulary you do, and they do not think about the user experience the way you and your colleagues do. You are too close to your product. You've assigned words and concepts to security and privacy that your users have not. Know that your users may not use the words "security" and "privacy" or even think about these concepts the same way you do. Understand their mental models and use their vocabulary—don't assign them your own.

Third, I was guilty of unintentionally priming participants during research studies, which biased my study results. *Priming* means that you say or do something during a research study that influences a user's behavior. Just saying words like *privacy*, *safety*, and *security* can prime participants. UC Berkeley and International Computer Science Institute (ICSI) researcher and founder of AppCensus Serge Egelman explains, "If you tell users you are studying security, they are more likely to pay more attention to security [than normal]."[3]

---

3   Serge Egelman, interviewed by Heidi Trost, December 20, 2023.

Imagine you are a research participant, and the moderator says, "The purpose of our study today is to evaluate whether you ignore or pay attention to a series of security warning messages. We're testing a few different variations to see if any of them cause our study participants to pause and take notice." Obviously, you are going to pay much more attention to those warning messages!

Often, in user research, we don't tell users everything about the study because we don't want to prime them. But what you don't tell participants—or the situations you place them in—could also be harmful to them. Be very mindful of not putting your research participants—or their information—at risk (read the following sidebar "Do Not Put Your Research Participants—or Their Information—at Risk").

So, what can you do? You don't want to prime participants or bias your research results. But you also don't want to harm your participants or put them at risk. The following recommendations, expanded on in the next few sections, can—when combined together—help:

- **Review, summarize, and share secondary research.** As a weekly practice, sharing and summarizing secondary research can help you stay updated on the latest security information, understand how your users might be put at risk and react, as well as serve as a way for your team to bond around security.

- **Gather security user experience data you already have.** Your existing research repository, customer support transcripts, and behavioral analytics often already have a wealth of information that can serve as a starting point or help you triangulate with other research you conduct.

- **Lean into the research you're already conducting.** Remember, you don't want to separate security and privacy from the rest of the user experience. Security should be studied in the context of a broader goal. Get in the habit of tagging insights related to the security user experience in all of the UX research you conduct.

NOTE LEAN ON YOUR UX RESEARCH TEAM

If your team includes UX researchers, this is their area of expertise. Hopefully they have already been included in your conversations so far. Lean on their experience and get them involved early. They can help you craft research questions, choose the right research methods, and guide you through getting the insights you need to move forward.

Let me be explicit: You should never put participants or their information at risk during research studies.

I also advise teams against leading their users to *believe* their information might be at risk during the study. This is known as *deception*. *Deception* in security UX research studies, as defined by usable privacy and security researchers,[4] means that you make the research participant believe their information is truly at risk and later, after the study is over, debrief the participant to explain that they (and their information) were not actually put at risk.

If you are contemplating this, first put yourself in the situation of your research participants. How would you react if put in the same situation? You might have a very negative, visceral reaction during the study. And you certainly would not trust the researcher once you realized the deception had taken place. Be very careful how you design and conduct the study to protect your participants.

Here is an example of how researchers tried to create a scenario to be as realistic as possible, without putting their participants at risk. Researchers Elaine Lau and Zachary Peterson, who were interested in how people who used screen readers experienced and reacted to browser security warnings like "This site might harm your computer," did a few things to ensure that they got the insights they needed without putting their research participants at risk.[5]

- **They were transparent about what they were researching upfront.**
- **They had participants visit test warning pages implemented by the most popular browsers.** This provided a realistic experience as participants used assistive technologies without putting participants—and their information and devices—in harm's way.
- **They made sure that participants knew they could skip a question or activity and end the session whenever they wanted to.**

Here are a few additional suggestions:

- **Anticipate where you might make participants uncomfortable.** Even recounting past experiences—an invasion of privacy, the feeling of losing

---

4   Verena Distler et al., "A Systematic Literature Review of Empirical Methods and Risk Representation in Usable Privacy and Security Research," *ACM Transactions on Computer-Human Interaction 28*, no. 6 (2021): 1–50, https://doi.org/10.1145/3469845

5   Elaine Lau and Zachary Peterson, "A Research Framework and Initial Study of Browser Security for the Visually Impaired," *Proceedings of 32nd USENIX Security Symposium* (USENIX Security 23), (2023), www.usenix.org/conference/usenixsecurity23/presentation/lau

control, or being scared for a loved one—can be potentially painful or traumatic to research participants. Don't let these situations surprise you—plan for them. Be really mindful of the situations you are putting your participants in and talk with your team ahead of time about how you will debrief participants to ensure that they feel safe and comfortable after their session is complete. There should be no question in your participant's mind that anything they did during the research would put them or their information at risk.

- **Make peer reviews a mandatory part of your UX research process.** Someone other than the team or person planning and conducting the research should review the research plan prior to the research starting. The review should focus on providing broader perspectives on research questions, overall approach, and methodologies, as well as feedback on interview or survey questions and task scenarios used in usability studies. The reviewer should question whether the study is ethical, where participant information may be put at risk, and most importantly, places where participants may feel uncomfortable or experience harm during or after the study.

- **Consult with your legal team about participant information you gather during the study and what type of consent you need to obtain.** Obtaining consent means participants know what they are signing up for. Inform your legal team if you plan to ask participants to install software or make any changes to their device, or if you may be exposed to customer personal data (even if inadvertently).

- **Have a process for how you collect, share, and dispose of personal information collected during UX research.** Keep in mind that the data you gather through your research may contain personal information— such as information on screenshots or screen recordings. When you are storing and circulating these research artifacts to different team members, you could inadvertently create a security risk. Plan ahead—and consult with your legal and privacy teams—on how you will store, share, and dispose of research artifacts that may contain personal information. Be transparent with participants about what you are collecting and how it will be shared—and stick to that promise.

- **Know when to move on.** During a research session, use your judgment. If you feel like you are making a participant particularly uncomfortable, skip to the next question. In some cases, you might offer to end the study early (but you should still give the participant their full incentive, regardless of whether you completed the study or not).

Review, Summarize, and Share Secondary Security UX Research

This might sound really obvious, but the best way to learn about and to be constantly thinking about the security user experience is to have it slow-drip into your cross-disciplinary team's consciousness.

Find research papers, listen to podcasts, and subscribe to newsletters. Then, every week, someone on your cross-disciplinary team should review, summarize, and share something that stood out to them.

Why is this helpful?

- It exposes you and members of your team to ideas, terms, and research methods outside of your feature, product, and organizational bubble.

- It builds up your shared knowledge quickly.

- It fosters cross-disciplinary bonding, as it often sparks questions and discussion.

- It helps your team stay up-to-date on constantly changing security threats. Having these things in the back of your mind helps you brainstorm more about "What can go wrong?" scenarios as you design new products and features.

- Learning about how others conduct research can illuminate possibilities for your own UX research.

Here's what makes this practice most successful:

- Post regularly, such as every Friday.

- Post in a place where your team will see it, such as in Slack or Teams.

- Rotate team members. Ideally, this is not just one person's responsibility or representing one person's interests. However, your team's lack of participation should not dissuade you from posting. In my experience, once you start sharing, others will join in naturally.

- Don't just post a link and say, "Hey team, check this out." Instead, summarize the key takeaways, how it applies to what you're working on, and include screenshots where possible. This is especially important for research papers, which often appear to be written to be intentionally difficult to read. Save your colleagues from this special kind of hell by putting the key takeaways in bullet points.

Gather Security User Experience Data You Already Have

There are several places where you may already have data around the security user experience. You can learn from and build upon this data:

- **Your existing UX research insights repository.** A research repository is simply a collection or database of research and research insights. Your team might use software designed for this purpose or just use a spreadsheet. You can search for specific words or phrases (for example, "sign in" or "two-factor authentication"), but you'll capture more insights if you start broadly and then narrow it down. If possible, focus on a specific user journey. Reference Chapter 1, "Security Impacts the User Experience," to think about where security is most likely to impact your users.

- **Talk to colleagues in customer or product support or in customer experience.** These people talk to your users every day. Understand what security-related user experience issues they encounter the most when working with customers. Fostering a good relationship and sharing information with these teams is a good practice all around. All of your jobs center around improving the customer experience—you should not be operating in silos.

- **Analyze support calls, chat recordings, or transcripts.** Your organization may already code the transcripts for support requests, making it easier to find calls or chats that mention something in a specific category (for example, "password").

- **Leverage behavioral analytics.** What percentage of your users are using the security feature? Are users on the page long enough to actually read the security-related instructions? What do they do next? Behavioral analytics have the benefit of helping you understand how users are using your product in the real world.

This information helps you triangulate data, leading to better insights.

Lean into the Research You're Already Conducting

You don't need to do security-specific UX research to gather security UX insights. Instead, pull out the security-specific insights from the research you're already conducting. Get in the habit of tagging insights related to the security user experience in all of the UX research you conduct. Put those in your research insights repository. And make an extra effort to share those insights with the people or teams they impact.

Think about the team designing the eero setup process, described in Chapter 5, "Design for Secure Outcomes." They probably went through several iterations of that onboarding flow, testing different options, reducing the number of steps, and rewording confusing instructions.

During those design iterations, they could conduct usability studies on different prototypes. But instead of telling users that security was one of the things they were researching, they'd simply observe and document whenever security or privacy came up as part of the study. During research studies, don't use security- or privacy-related words or terms your users don't say. Don't assume understanding where there is none.

For example, a participant may explicitly say something like, "I don't know why it's asking me for this." Say, "tell me more" and listen (also a tried-and-true Nielsen Norman Group recommendation). Or you might notice nonverbal cues indicating a participant is confused or uncomfortable. Lean into those moments. "Tell me what is going through your mind there."[6]

They might say, "I have no idea what WPA3 means." Yes, this is a product setup insight, but it's *also* a security-related insight—information the team can use to improve the setup *and* security user experience.

Include Users with Disabilities in Your Research

Make sure that you include users with disabilities in your research. Maya Alvarado, senior accessibility researcher at Booking.com, says, "Recognize that accessibility research is still user research, but just more inclusive. We should build for flexibility in timings and methods of collecting feedback. Let participants decide what is most comfortable for them, e.g., written responses or recording a video."[7]

---

6   The seminal work on how to prepare for and conduct user interviews is Steve Portigal, *Interviewing Users*, 2nd ed. (New York: Rosenfeld Media, 2023).

7   Maya Alvarado, "Building Accessibility Research Practices," Booking.com UX Research (article), October 24, 2023, https://medium.com/booking-research/building-accessibility-research-practices-75d82098f286

You can approach this research as a co-creation exercise—not something you "check off" as part of your UX research. Accessibility trainer and educator at VMware and accessibility advisor for the FIDO Alliance task force Joyce Oshita stresses, "You have to have people with disabilities co-creating with you [...] How can you meet with me for an hour and know how I operate?"

Joyce, who "navigated slow vision loss for decades before it transitioned into blindness," is working on a video series called *Digital Overload*, which documents her experience using different digital services using a screen reader.[8] In "A Blind User's Encounter with 'Accessible' CAPTCHA," Joyce is prevented from completing a task due to security-related roadblocks.[9] In this particular video, not only is the audio version of the CAPTCHA hard to understand in general, Joyce's screen reader is competing with the audio, making it incredibly difficult to know what has been entered and what part of the CAPTCHA still needs to be completed. If your team does not yet include people with disabilities as part of your research, I recommend checking out and sharing Joyce's videos as part of making your case for doing so.

I also highly recommend visiting "Microsoft's Inclusive Design" website and incorporating their resources and toolkits into your design and research practice. Their "Inclusive 101 Guidebook"[10] will help your team brainstorm how permanent, temporary, and situational disabilities may impact people using your product.

In addition, the "Inclusive Design for Cognition Guidebook"[11] is an excellent resource to better understand how people might perceive and understand security- and privacy-related communications, tasks, and warnings. When the security user experience fails to take into account cognitive demands—forcing users to read, understand, and act on a complex set of instructions or interact with several services,

---

8   Joyce expands on why she started the *Digital Overload* series: Voice+Code, "Include Users with Disabilities in Your Security UX Research with Joyce Oshita," *Human-Centered Security* (podcast), May 22, 2024, https://share .transistor.fm/s/cfbdc122

9   Joyce Oshita, "E1—A Blind User's Encounter with 'Accessible' CAPTCHA," *Digital Overload* (video), www.youtube.com/watch?v=8ttExPtn2iE

10 "Inclusive 101 Guidebook," Microsoft (guide), https://inclusive.microsoft .design/tools-and-activities/Inclusive101Guidebook.pdf

11 "Inclusive Design for Cognition Guidebook," Microsoft (guidebook), https://inclusive.microsoft.design/tools-and-activities/InclusiveDesignFor CognitionGuidebook.pdf

possibly using different devices—you are significantly reducing the chances that users will take the safer path.

If you don't include people with disabilities in your research, you may not realize design choices you've made could negatively impact that person's security or privacy. For example, take the "Deceptive Site Ahead" warning described in Chapter 1. Researchers Elaine Lau and Zachary Peterson (whose research was referenced earlier in this chapter) explain, "While navigating through site content and meta information, visually impaired users have to split their cognitive energy in three ways between interpreting the website contents, screen reader, and browser."[12]

Warnings are often dense with information, and consequently, someone using a screen reader is also relying on text descriptions of any visuals (like icons). The researchers found if those text descriptions were overly long and unhelpful, they could actually get in the way of the intended message. As a result, the researchers warned that users using screen readers may actually be less likely to heed the warning.

Additional places to learn more:

- **WCAG guidelines** (version 2.2 as of the writing of this book): Guidelines on how to make digital experiences more accessible.[13]
- **Accessible security design research papers:** Research papers are helpful in not only learning about the latest research being done in accessible security, but also in helping your team learn about how to design accessible security research studies. While not focused exclusively on accessibility, I recommend you check out conference proceedings for USENIX Symposium on Usable Privacy and Security (SOUPS) and search for papers from researchers who have included people with disabilities as part of their research.

---

12 Lau and Peterson, "A Research Framework and Initial Study of Browser Security for the Visually Impaired."

13 "Web Content Accessibility Guidelines (WCAG, 2.2)," October 5, 2023, www.w3.org/TR/2023/REC-WCAG22-20231005/

Align on Goals and Define Security UX Metrics to Help You Iterate

Here's something that is really important: you cannot measure anything that will meaningfully inform your feedback loop—and help you make design changes—if you don't know what your goals are. Worse, not having a goal is a huge waste of time.

As suggested in Chapter 1, define words like *trust*, *security*, and *privacy* with your cross-disciplinary team. You can't possibly have goals until you define what these words mean in the context of using your product. What are you trying to achieve? Is that the same thing your users want? How do you know?

I can tell you from experience that your users are asking the same thing. In other words, they are asking, "What do you mean you take my privacy seriously?"

One piece of advice: Make sure that everyone, including your security UX allies, are part of defining what these goals are. Have you asked your security UX allies what they want to measure? Did your UX changes contribute to making the user or system more secure (again, according to your shared definition of what *secure* means)?

> **NOTE** **"BUT WE'LL NEVER GET BUY-IN"**
>
> I've had many people say, "Investing in security UX sounds great, but we'll never get buy in." Then stop using the word *security* and focus instead on *trust*. *Trust*—or *lack of trust*—is where you gain or lose customers. In other words, trust is where the business makes money or loses money.
>
> If people don't trust you with their information, they won't sign up. If they lose trust in you by the way you (mis)handle their information, they'll leave. If they can't sign into their account, they'll leave (or rather, ironically, you've made it so they can't come back). Not to mention the thousands of confused and angry customer service messages you'll receive.
>
> Many of these moments are places in Alice's user journey that you learned about in Chapter 1. Focus on these places first.

Working closely with your security UX allies on defining goals and metrics is a win-win. You're looking to demonstrate a return on investment, and they have business goals they are trying to meet.

Further, they may be able to help you set up or get access to the most effective measurement tools you'll need.

Once you define what it means, how do you know that you've reached that goal? If your team is new to UX metrics, I strongly recommend reading about Google's HEART Framework, which helps your team break down how to move from a goal to a UX metric. (HEART stands for *happiness, engagement, adoption, retention,* and *task success.*[14]) I've found that leveraging the Google HEART framework is helpful in getting teams to think through user experience metrics.

Also check out "Measure What Matters: Crafting UX Success Metrics" a recording of a workshop given by Kate Rutter. This is one of the best resources to learn how to define UX metrics.[15]

---

14 Kerry Rodden, Hilary Hutchison, and Xin Fu, "Measuring the User Experience on a Large Scale: User-Centered Metrics for Web Applications," *Proceedings of the SIGCHI Conference on Human Factors in Computing Systems* (April 2010): 2395–2398, https://doi.org/10.1145/1753326.1753687

15 Kate Rutter, "Measure What Matters: Crafting UX Success Metrics," workshop recording for True Ventures, June 13, 2017, https://youtube.com/watch?v=x4T_Eg46L4c

This chapter only scratches the surface of UX research. Here are general UX research resources (not security-specific):

- *Build Better Products* by Laura Klein[16]
- *Just Enough Research* by Erika Hall[17]
- *Interviewing Users*, 2nd ed. by Steve Portigal[18]
- "A Guide to Using User-Experience Research Methods" by Christian Rohrer and Kelly Gordon[19]
- "When to Use Which User Experience Research Methods" by Christian Rohrer[20]
- Steve Krug's *Rocket Surgery Made Easy*[21]

Here are security-specific UX research resources:

- "A Systematic Literature Review of Empirical Methods and Risk Representation in Usable Privacy and Security Research"[22]
- CyLab Usable Privacy and Security Laboratory at Carnegie Mellon University, which has a list of usable privacy and research studies[23]
- USENIX Symposium on Usable Privacy and Security (SOUPS)

16 Laura Klein, *Build Better Products* (New York: Rosenfeld Media, 2016).

17 Erika Hall, *Just Enough Research*, 2nd ed. (New York: A Book Apart, 2019).

18 Portigal, *Interviewing Users*, 2nd ed.

19 Kelly Gordon and Christian Rohrer, "A Guide to Using User-Experience Research Methods," Nielsen Norman Group (article), August 21, 2022, www.nngroup.com/articles/guide-ux-research-methods/

20 Christian Rohrer, "When to Use Which User Experience Research Methods," Nielsen Norman Group (article), July 17, 2022, www.nngroup.com/articles/which-ux-research-methods/

21 Steve Krug, *Rocket Surgery Made Easy* (Berkeley, CA: New Riders, 2009).

22 Distler et al., "A Systematic Literature Review of Empirical Methods and Risk Representation in Usable Privacy and Security Research."

23 https://cups.cs.cmu.edu

Shift Your Mindset

In this chapter, you learned:

- You won't be able to anticipate every possible scenario in the security user experience. That's why having an iterative mindset both during the design phase and post-launch is so important.

- In order to improve the security user experience, you have to understand for and account for the *entire* user experience. Be careful about isolating security from the rest of Alice's experience.

- Be careful not to prime participants in user research studies and be very mindful about never putting them or their information in harm's way.

- Lean on UX research you're already conducting—you're likely capturing useful security- and privacy-related information. This should be tagged in your research repository.

- Include people with disabilities in your research. But approach this research as a co-creation exercise—not just something you "check off" as part of your UX research.

- Make sure that your team is aligned on your goals. Figure out what security UX metrics might make sense based on your goals.

Your Users Are Relying on You

You've probably watched a toddler grab their parent's iPhone, effortlessly enter the passcode, and navigate to their favorite game or YouTube video. Many little ones who can barely form full sentences can find a new game on the App Store and place a FaceTime call. I recently observed a toddler tap a big screen television, expecting it to respond like a tablet. (He was befuddled when it did not comply.)

Toddlers can learn how to use an iPhone so quickly not just because they are quick studies and mimic their parents, but also because the design is intuitive. You, UX designers, designed these experiences.

But when you watch a toddler—or even a teenager, your partner, or your parents—using a smartphone, they often aren't thinking about security. That smartphone is an extension of themselves.

Soon that toddler is going to be using an Apple Vision Pro just as effortlessly, and he's going to look at your old smartphone like it's a dinosaur. His digital and physical worlds are going to merge even more than yours have already. And he's going to feel sorry that you had to live part of your life without a personal AI assistant that helps you manage your life and make better decisions.

Soon, there will be services that you will have no other choice other than to access digitally.

When that toddler—Alice—grows up and uses whatever gadget is popular in the future, she doesn't want to worry about security. She wants to get things done, be entertained, and live her life.

In fact, as you learned in Chapter 1, "Security Impacts the User Experience," Alice expects her experience to be safe. Rachel Botsman, in her book *Who Can You Trust?*[1] defines trust as "a confident relationship with the unknown." There's a lot of trust that goes into using technology—technology most users know very little about. Users are putting their confidence in you, as people who design and build products, that things will go as expected. And baked into those expectations, often, is that the experiences users have with your product are safe.

---

1   Rachel Botsman, *Who Can You Trust? How Technology Brought Us Together and Why It Might Drive Us Apart* (New York: PublicAffairs, 2017), 20.

Keep in mind the quote from Kelly Shortridge (from Chapter 1), "We should aim to provide subjective security, that ancient-school version of *securitas*, which meant freedom from anxiety, fear, or care."[2]

Alice needs your help to have a safe experience. But you can't possibly help her unless two things happen first:

1. You have to understand the security UX ecosystem.
2. You have to work with your security UX allies to complete your understanding of the security ecosystem and, in turn, help them understand Alice better.

How? Tell better stories.

Tell Better Stories

Stories—how Alice, Charlie, and the threat actor all interact and influence each other in the security ecosystem—will help you visualize and better understand what can often be complicated, ever-changing dynamics.

Then you can use this understanding to explain it to your security UX allies. But remember, you have an incomplete story. Your security UX allies can help you fill in the missing pieces and edit that story as it evolves (because it will!). As a result, your—and your colleagues'—understanding will deepen.

When Christian Rohrer (referenced in Chapter 1) introduced the idea of the three players in the security ecosystem, I felt like I finally had a story to tell, and it went something like this:

> Alice is your user. She's human, and she does human things. She's trying to balance all the things. She's doing her best.
>
> There are threat actors ready to take advantage of Alice. They know she's human. They know she makes mistakes. They know how to trick and manipulate her.
>
> It's not fair.

---

2 Kelly Shortridge, "When We Say 'Security,' What Do We Mean?," *kellyshortridge.com* (blog), October 26, 2023, kellyshortridge.com/blog/posts/what-does-the-word-security-mean

But what's worse is the moments where Charlie might be able to help Alice, but he doesn't. He lets her figure things out herself. He confuses her by using words she doesn't understand. He fails to anticipate where she might forget or do something wrong.

This seems *really* unfair.

The good news is that you can help.

First, find your security UX allies. Next, ask the right questions—a good place to start—see Chapter 4, "Find the Right People, Ask the Right Questions"). Finally, work with your security UX allies to understand and illustrate how these dynamics play out. As someone in UX, you are in the best position to tell the story of Alice, Charlie, and the threat actor (see Figure 8.1).

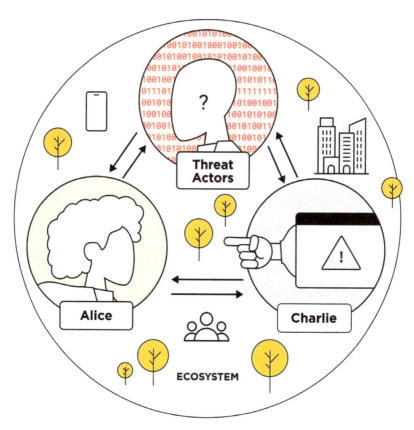

FIGURE 8.1
Everything in the ecosystem affects everything else.

Now bring that story to life for *your* Alice, *your* Charlie, and the threat actors that are unfortunately part of your ecosystem.

When you better understand how the story is playing out, you can work toward a happier ending.

You've made products that are easy to use. Now you—with the help of your security UX allies—need to help secure them.

Help keep people safe. And, hey, stay secure.

Shift Your Mindset

In this chapter, you learned:

- The analog and digital worlds have already collided. Hopefully, you already knew that.
- You've designed some of the most amazing digital experiences. You need to help secure those experiences.
- You have to understand the security ecosystem. Then you have to work with your security UX allies to complete your understanding of the security ecosystem and, in turn, help them understand Alice better.
- You can do this by telling better stories. Help everyone on your team understand the stories around Alice, Charlie, and the threat actor.

INDEX

K

Keenoy, Tom, 77–78
knowledge-based mistakes, 35
Kohnfelder, Loren, 54

L

lapses (security mistakes), 37, 77
Lau, Elaine, 114, 120
leadership of company, 28, 51–52
learning from users. *See* user experience
  (UX) research
legal team, 28, 51–52, 115
locked front door, 105

M

Mailchimp, 81–82
"Measure What Matters" (Rutter), 122
metrics, and UX research, 121–122
Metts, Michael J., 84
Microsoft, Inclusive Design resources,
  104, 119, 120
mistakes, as security incidents, 34–37, 77
MITRE ATT&CK framework, 37
modems, 75–76
Mozilla Firefox, warnings, 14
multifactor authentication (MFA), 12,
  43–45, 58, 97
  MFA fatigue, 46, 47

N

National Institute of Standards and
  Technology (NIST), "Guide for
  Conducting Risk Assessments," 19
Nielsen, Jakob, 88
Nobles, Calvin, 101
Norman, Don, 74
Nudge: The Final Edition (Thaler and
  Sunstein), 9, 72, 89
nudges, 9

O

onboarding, 62–63, 75–78
one-time passcode (OTP), 77
Oshita, Joyce, 119

P

passkeys, 96–97, 108–111
passwords
  as authentication factor, 96
  in compromised accounts, 56–59
  the need for another factor, 44–45
  passkeys as alternative, 108–109
  for routers, 76–77
peer reviews in UX research, 115
permissions and authorization, 95
personal information
  processing of, defined, 20
  in UX research, 115
  what needs to be protected, 23
  where security impacts the user
    experience, 13
Peterson, Zachary, 114, 120
principle of least privilege, 95–96
"Principles of Human-Centered
  Design" (Norman), 74
priorities, competing, 27–29, 30–31
privacy, defined, 20
The Privacy Engineer's Manifesto
  (Dennedy, Fox, and Finneran), 20
privacy settings, 13–16
privileged access, 64–65
privileges and authorization, 95
product managers, 28, 51–52
Puglisi, Jason, 74
push notifications, 45–46
  modified, 47

ACKNOWLEDGMENTS

This book truly took a village to write. So many people offered to help by sharing their expertise and providing feedback. I learned so much from you. I appreciate you all!

Thank you to my mom and G. You inspired me to better understand the "human" in human-centered security. And you made sure I was periodically fed, watered, and given hugs.

Thank you Christian Rohrer, a UX leader who was practicing human-centered security long before I started thinking about the topic. Christian inspired this book with his description of the different players in the security ecosystem. He also provided invaluable feedback throughout the writing of the book.

Thank you Dave Johnson for encouraging me to tell my story—for giving me confidence that as a UX researcher I could have something meaningful to say about security. Dave acted as a security yoda, editor, mentor, and therapist all in one package. I could not have finished the book without him.

Thank you Lou Rosenfeld for giving me the opportunity to write about the security user experience and Marta Justak for patiently helping me through countless iterations of the book.

Thank you to the people who graciously offered their insights, perspectives, and feedback on the book:

Matthew Bernius, Keelin Billue, Abby Bridges, Sheri Byrne-Haber, Lorrie Cranor, John Crowley, Michelle Finneran Dennedy, Terri Ducay, Carl Edholm, Pat Forbes, Gabriel Friedlander, Kevin Goldman, Kendra Graham, Jay Harlow, Devon Hirth, Carlie Hundt, Liya James, Tom Keenoy, Brandie Knox, Radhika Koyawala, Sandy Martinuk, Calvin Nobles, Meghan O'Meara, Joyce Oshita, Jason Puglisi, Jarret Raim, John Robertson, Nikki Robinson, Paul Seltmann, Kristen Seversky, Adam Shostack, Michael Snell, Paul Solt, Bethany Sonefeld, Jeremiah Still, Wako Takayama, Jason Telner, Paul van Oorschot, Ami Walsh, Andrew Ward, Amy Wells, Ira Winkler, James Wondrack, Kaliya Young.

 **Rosenfeld**

Dear Reader,

Thanks very much for purchasing this book. There's a story behind it and every product we create at Rosenfeld Media.

Since the early 1990s, I've been a User Experience consultant, conference presenter, workshop instructor, and author. (I'm probably best-known for having cowritten *Information Architecture for the Web and Beyond*.) In each of these roles, I've been frustrated by the missed opportunities to apply UX principles and practices.

I started Rosenfeld Media in 2005 with the goal of publishing books whose design and development showed that a publisher could practice what it preached. Since then, we've expanded into producing industry-leading conferences and workshops. In all cases, UX has helped us create better, more successful products—just as you would expect. From employing user research to drive the design of our books and conference programs, to working closely with our conference speakers on their talks, to caring deeply about customer service, we practice what we preach every day.

Please visit **rosenfeldmedia.com** to learn more about our **conferences**, **workshops**, **free communities**, and **other great resources** that we've made for you. And send your ideas, suggestions, and concerns my way: louis@rosenfeldmedia.com

I'd love to hear from you, and I hope you enjoy the book!

Lou Rosenfeld,
Publisher

RECENT TITLES FROM ROSENFELD MEDIA

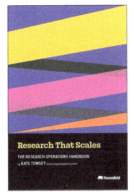

Get a great discount on a Rosenfeld Media book:
visit rfld.me/deal to learn more.

SELECTED TITLES FROM ROSENFELD MEDIA

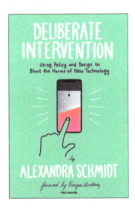

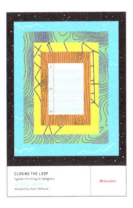

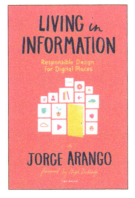

View our full catalog at rosenfeldmedia.com/books

ABOUT THE AUTHOR

Heidi Trost is a UX leader who helps cross-disciplinary teams improve the security user experience. With a background in UX research, Heidi does this by helping teams better understand the people they are designing for, as well as the security threats that may impact people and systems negatively. Heidi is also the host of the podcast, *Human-Centered Security*, where she interviews security experts and people who design for the security user experience. When not thinking about security, you can find her in a sunny spot reading a book, hiking, or riding horses.

www.ingramcontent.com/pod-product-compliance
Lightning Source LLC
LaVergne TN
LVHW011803070326
832902LV00026B/4620